GUIDELINES FOR DETERMINING THE PROBABILITY OF IGNITION OF A RELEASED FLAMMABLE MASS

This book is one in a series of process safety guideline and concept books published by the Center for Chemical Process Safety (CCPS). Please go to *www.wiley.com/go/ccps* for a full list of titles in this series.

GUIDELINES FOR DETERMINING THE PROBABILITY OF IGNITION OF A RELEASED FLAMMABLE MASS

Center for Chemical Process Safety
New York, NY

WILEY

Library of Congress Cataloging-in-Publication Data:

Guidelines for determining the probability of ignition of a released flammable mass / Center for Chemical Process Safety, New York, NY.
    pages cm
  Includes bibliographical references and index.
  ISBN 978-1-118-23053-4 (cloth)
1. Flammable materials—Fires and fire prevention—Mathematical models. 2. Chemicals—Fires and fire prevention—Mathematical models. 3. Chemical processes—Safety measures—Mathematical models. 4. Probabilities—Mathematical models. I. American Institute of Chemical Engineers. Center for Chemical Process Safety.
  TH9446.I47G85 2014
  660'.2961—dc23   2013042730

10 9 8 7 6 5 4 3 2 1

CONTENTS

LIST OF FIGURES

LIST OF TABLES

FOREWORD

When making risk estimations, two variables are involved; the probability that an event will occur and the consequence, the logical and expected result of that event. For unplanned events these are independent of each other. Of the two, more development and investigative time has been spent on determining consequence. Multiple methodologies and approaches are available to reasonably predict the consequences of a chemical release, fire, and/or explosion on surrounding manufacturing equipment, people, the environment, etc. The technology to do so is well developed and is enhanced when new information becomes available. Therefore, consequence estimation is not within the scope of this book.

This book focuses on probability, which is more difficult to quantify objectively or experimentally. With the help of member companies of the Center for Chemical Process Safety (CCPS), research for the book began with a co-op student directed and assisted by a professional librarian and supplemented by a subcommittee composed of volunteers from CCPS member companies.

After approximately two years of research and document reviews, the subcommittee agreed that there was sufficient information and technology available to produce a book that addressed and reasonably estimated the component of the risk equation relating to the probability that a released flammable mass would ignite. Available technology methods could be compiled, codified, and developed into real-world usable tools as part of the book. The CCPS membership approved the proposal to write a book on the subject and a book subcommittee was formed.

The information used to write this book and develop the calculation tool came from the previous research conducted which is not proprietary and is generally available in the public domain. The CCPS subcommittee compiled, codified, clarified, and in some cases further developed the information and data to develop the algorithms in the book and the associated calculation tool. This tool, when populated with appropriate data, yields an estimate of the probability that a defined release of a flammable material will ignite if exposed to an ignition source. This tool can provide users, within reason, information that can be used in risk assessments with a higher degree of confidence than estimates made before.

This book and calculation tool have been developed and are presented with the understanding that users will employ whatever methodology they deem appropriate to estimate the probability that the released flammable mass will reach an ignition source. That estimate is then combined with the probability of ignition determined per the methodology in this book and the associated calculation "tool" to determine the total estimated probability of ignition of a release.

The team agreed that the calculation tool should be conservative, especially when data was scant and/or known to be beyond previously vetted values. The team believed it would be safer and more responsible to estimate a higher probability of ignition than to rely on unconfirmed or proprietary data and methodologies or approaches designed for very specific uses. The desire to err on the side of being conservative was tempered by the committee's recognition that being "conservative" could lead to other issues, e.g., predicting a high ignition probability reduces the probability of a toxic outcome for chemicals that are both flammable and toxic. Additionally, the team recognized that some companies may have proprietary or more detailed data than what is publicly available and could reasonably reach a different estimate as to the probability that a defined release of a flammable material would ignite when exposed to an ignition source. It is critical that users thoroughly read the book to understand the scope and limitations of both the book and calculation tool. The calculation tool is populated with popup caution "flags" to warn the user when the calculation tool is being used at the extremes of its design capabilities as we know them now. However, not all such conditions may be identified at present.

Readers should be aware that the methodologies presented here apply only to process safety incidents, unplanned events where probability and consequence are independent of each other. These methodologies may not be used in evaluating the risk associated with willful acts such as sabotage or terrorism. In such cases, probability and consequence are codependent, i.e., consequence influences probability.

Like many CCPS books, this book breaks new ground. As such, we anticipate that over time these methodologies will improve. For this we would like to enlist your help to improve the precision and utility of the calculation tool. We intend to provide periodic updates to the tool and within a few years determine if a second edition of the book and an updated version of the tool are warranted. The CCPS website provides a way for users to submit information and lessons learned related to usage of the tool and a way for users to download new information and updated versions of the tool should there be any.

Owners of this book can access the calculation tool on the CCPS website:

http://www.aiche.org/ccps/resources/publications

This "tool" is provided as part of the book purchased and should not be used by anyone other than the owner of this book. The tool should also only be used after the user has read the book, particularly Chapters 1 and 4, and should initially be used in conjunction with the illustrations in Chapter 5.

User Please Note: This tool represents technology that is evolving and is based on the algorithms and logic described in the corresponding CCPS book, "Guidelines for Determining the Probability of Ignition of a Released Flammable Mass." Any such tool cannot anticipate all possible circumstances of usage, and it is expected that the user of this tool has read and understood the scope of the associated book, the tool, and the potential hazards associated with misuse of the tool. Results generated by users who have not read the book may in certain circumstances be incorrect or misinterpreted. Misunderstanding of how to use the tool and/or misuse of the tool may result in inaccurate reported results and possibly inappropriate actions on the user's part.

It is sincerely hoped that the information presented in this tool will lead to an even more impressive safety record for the entire industry; however, the American Institute of Chemical Engineers (AIChE), its consultants, the AIChE's CCPS Technical Steering Committee and the Probability of Ignition subcommittee members, their employers, their employers' officers and directors, and Baker Engineering and Risk Consultants, Inc. (BakerRisk) and its employees, officers, and directors neither warrant nor represent, expressly or by implication, the correctness or accuracy of the content of the information presented in this tool. As between (1) the AIChE, its consultants, the CCPS Technical Steering Committee and Subcommittee members, their employers, their employers' officers and directors, and BakerRisk, and its employees, officers, and directors and (2) the user of this tool, the user accepts any and all legal liability and responsibility whatsoever for the consequence of its use or misuse of this tool.

ACKNOWLEDGMENTS

The American Institute of Chemical Engineers (AIChE) and its Center for Chemical Process Safety (CCPS) express their appreciation and gratitude to all members of the Probability of Ignition project subcommittee and their CCPS member companies for their generous support and technical contributions in the preparation of these guidelines.

SUBCOMMITTEE MEMBERS:

Subcommittee Members:

Robert Stack—Chair	The Dow Chemical Company
John Baik	BP
Laurence G. Britton	Expert & Process Safety Consultant
Mervyn Carneiro	Eli Lilly and Company
Wayne Chastain	Eastman Chemical Company
Andrew Crerand	Shell Projects and Technology
Trond Elvehoy	DNV
Jeffrey Fox	Dow Corning Corporation
Randy Hawkins	AON Energy Risk Engineering
David Herrmann	DuPont
Jack Reisdorf	Fluor
Jim Salter	Chevron
Vince Van Brunt	University of South Carolina

Principal Author: Michael Moosemiller, Baker Engineering and Risk Consultants, Inc.

CCPS Staff Consultant: Adrian L. Sepeda , Center for Chemical Process Safety

CCPS acknowledges the many contributions of the principal author, Mr. Mike Moosemiller, who went well beyond the required to accommodate the schedule of the subcommittee, the numerous reviews and edits and uncountable number of trials and vetting of the calculation software "Tool." CCPS also recognizes the continuous and guiding efforts of the subcommittee chair, Mr. Bob Stack, who guided the effort through this long and difficult task with continuous support, encouragement, and unwavering dedication and direction.

CCPS thanks Baker Engineering and Risk Consultants Inc. ("BakerRisk") and all of their contributors that made the publication of this book possible:

Moira Woodhouse: Technical editing—BakerRisk

Jef Rowley: Writing code for software "Tool"—BakerRisk

Jesse Calderon: Writing code for software "Tool"—BakerRisk

Before publication, all CCPS books go through a thorough peer review process. CCPS gratefully acknowledges the thoughtful comments and suggestions of the peer reviewers. Their review and suggestions enhanced the accuracy and clarity of these guidelines and the associated calculation "Tool."

Although peer reviewers have provided many constructive comments and suggestions, they were not asked to endorse this book and were not shown the final draft before its release.

Peer Reviewers:

John Alderman   Hazard & Risk Analysis

Don Connolley   BP

Chris Devlin   Celanese

Kieran Glynn   BP

Bob Johnson   UNWIN Integrated Risk Management

Beth Lutostansky   Air Products and Chemicals, Inc.

Kimberly Mullins   Praxair

John Murphy   CCPS

Phil Partridge   The Dow Chemical Company (Retired)

Erick Peterson   MMI Engineering (Tool only)

Robin Pitblado   DNV

And thank you to Ms. Jamie Gomez, the co-op student who conducted the initial literature search confirming that there was sufficient information available to start this project.

GLOSSARY

Consequence: The potential effects of an explosion, fire, or toxic material release. Consequence descriptions may be qualitative or quantitative. The undesirable result of an incident, usually measured in health and safety effects, environmental impacts, loss of property, and business interruption costs.

Early Ignition: Synonymous with "immediate ignition" as used in this book— that is, ignition that occurs closely enough to the source of a flammable release in space and time such that the possibility of a flammable cloud sufficient to create a vapor cloud explosion is precluded.

Dispersion Model: Mathematical model that characterizes the transport of toxic/flammable materials released to the air

Flammable: Two definitions based on NFPA:

(a) per NFPA 30: a liquid with a closed-cup flash point below 100°F (37.8°C) and Reid vapor pressures not exceeding 40 psia at 100°F (37.8°C). Flammable liquids are called Class I liquids and have three subdivisions: Class IA: Those having flash points below 73°F (22.8°C) and boiling points below 100°F (37.8°C). Class IB: Those having flash points below 73°F (22.8°C) and boiling points at or above 100°F (37.8°C). Class IC: Those having flash points at or above 73°F (22.8°C) but below 100°F (37.8°C); or

(b) per NFPA 55: A gas that can burn with a flame if mixed with a gaseous oxidizer such as air or chlorine and then ignited. The term *flammable gas* includes vapors from flammable or combustible liquids above their flash points.

FMECA Failure Mode, Effects, and Criticality Analysis

Frequency: Number of occurrences of an event per unit of time.

Fundamental Burning Velocity: The rate of flame propagation relative to unburned gas ahead of the flame front; the burning velocity of a laminar flame under stated conditions of composition, temperature, and pressure in the unburned gas.

IDLH   "Immediately Dangerous to Life or Health," as described by the U.S. National Institute for Occupational Safety and Health.

Ignition: See Section 1.3.1.1 of book.

Immediate Ignition: Synonymous with "early ignition" as used in this book; that is, ignition that occurs closely enough to the source of a flammable release in space and time such that the possibility of a flammable cloud sufficient to create a vapor cloud explosion is precluded.

Impact: The results of a given consequence on the occupants of a building (i.e., occupant vulnerability).

Incident: An unplanned event with the potential for undesirable consequences.

Layers of Protection Analysis (LOPA): A process (method, system) for evaluating the effectiveness of independent protection layer(s) in reducing the likelihood or severity of an undesirable event.

Likelihood: A measure of the expected probability or frequency of an event's occurrence.

LFL (Lower Flammability Limit): The concentration of a combustible material in air below which ignition will not occur. It is often referred to as the Lower Explosive Limit (LEL). Mixtures below this limit are said to be "too lean."

Probability: The expression for the likelihood of occurrence of an event or an event sequence during an interval of time, or the likelihood of the success or failure of an event on test or on demand. By definition, probability must be expressed as a number ranging from 0 to 1.

Process Hazards Analysis (PHA): An organized effort to identify and evaluate hazards associated with chemical processes and operations to enable their control. This review normally involves the use of qualitative techniques to identify and assess the significance of hazards. Conclusions and appropriate recommendations are developed. Occasionally, quantitative methods are used to help prioritize risk reduction.

Qualitative: Based primarily on description and comparison using historical experience and engineering judgment, with little quantification of the hazards, consequences, likelihood, or level of risk.

QRA (Quantitative Risk Assessment):   The systematic development of numerical estimates of the expected frequency and/or consequence of potential accidents associated with a facility or operation based on engineering evaluation and mathematical techniques.

Risk-Based Inspection: A risk assessment and management process that is focused on loss of containment of pressurized equipment in processing facilities, due to material deterioration. These risks are managed primarily through equipment inspection.

Scenario: An unplanned event or incident sequence that results in a loss event and its associated impacts, including the success or failure of safeguards involved in the incident sequence.

Semiquantitative: Risk analysis methodology that includes some degree of quantification of consequence, likelihood, and/or risk level.

1 INTRODUCTION

1.1 OBJECTIVES

The goal of this book is to provide information and methods that can be used to estimate the probability of ignition for flammable gas and liquid releases to the external environment. This book and the accompanying software tools discuss technical material that the user should be familiar with prior to use. This book is intended for an audience of engineers and/or scientists who have experience with process safety and risk management systems.

The algorithms that are developed in this book are presented at different levels of sophistication to accommodate a wide range of users, including people in a process hazard analysis team who want an objective but crude prediction for risk ranking purposes or people performing quantitative risk assessments and developing relatively complex risk mitigation plans. Users can adopt the level of complexity and accuracy needed for their particular application with a commensurate level of effort in data input.

The scope of this book is limited to flammable gases, mists, and liquids. The designed application is for onshore facilities, although it may be possible to extend it to offshore applications if the user is able to properly account for the inherent differences between the two settings. This book specifically excludes the treatment of ignitable dusts for various reasons, not the least of which are: (a) the magnitude and physicochemical characteristics of dust clouds are very difficult to quantify for a given situation, particularly for dust "disturbance" events (in which accumulated dust dislodges from the tops of equipment and support structures) and (b) ignition probability data for dust ignitions are very limited at this time.

1.2 MOTIVATION FOR THIS BOOK

Up until the 1990s, many companies maintained groups of process safety specialists whose experience and expertise in different areas allowed in-house problem solving. Often, companies not only maintained safety test laboratories but performed safety research as well. Unfortunately, as safety technology has advanced it has become more complicated and difficult for most companies to apply. This book is intended to assist in-house risk analysts in one of the most difficult areas—estimating the probability of ignition of a given vapor cloud.

The motivation for this book is to achieve the following three primary outcomes:

- A standardized methodology for estimating probability of ignition that is open-source and can be applied consistently across the process industry

- Methods and tools that allow a user to estimate ignition probability quickly
- Ability to account for mitigation measures that reduce ignition probability

On the last bullet above, it is desired that a tool be able to address as many of the elements of the "fire triangle" as possible. In fact, the methods can address all sides of the triangle to varying degrees, but none completely, and all resulting in reductions in ignition probability rather than elimination of ignition altogether.

1.2.1   A Brief History of Fire Protection

Many catastrophic accidents in the process industries have resulted from the ignition of a flammable mass that was released into the environment. For this reason, safety professionals and regulators have continually sought methods to reduce the frequency of such events, and various approaches have been undertaken to accomplish this. Before the implementation of industry standards and codes, professionals used their individual and/or collective knowledge of past events and fire fundamentals to mitigate such events. Even in ancient Rome, the Emperor Nero developed regulations for fire protection after the city burned in A.D. 64. The Roman regulations included requirements for fire-resistant building materials and the use of separation distances, concepts that are still in use today.

The evolution from this knowledge-based approach into a series of industry-driven standards and codes occurred in order to share knowledge of flammable hazard management and to introduce standardized methods for dealing with flammable hazards. Not surprisingly, the nascent insurance industry of the nineteenth century promoted this initial effort, and various professional organizations were created in the twentieth century such as the National Fire Protection Association (NFPA), Society of Fire Protection Engineers (SFPE), and others in the U.S. and overseas. These organizations were instrumental in developing the field of flammables management.

The science of ignitions in the petroleum, chemical, and other industries developed in parallel. Klinkenberg and van der Minne (1958) provide references on static electricity in the industry that date back to the 1910s. The U.S. Bureau of Mines had a leading role in progressing knowledge in this area in the same time frame. Through these efforts and contributions by others in industry, advancements in both the theory and experimental support for these phenomena were made through the middle of the twentieth century.

As the chemical and petrochemical industries matured and grew, the potential for fires and explosions of ever-greater magnitudes also grew, and some tragic events such as those in Flixborough, Piper Alpha, Mexico City, and Pasadena drove regulators to become more intimately involved in the management of flammable hazards. In the U.S., the promulgation of the Occupational Safety and Health Administration's "Process Safety Management of Highly Hazardous Chemicals" standard in 1992 set the stage for the regulation of such hazards, although the standard is largely built on and refers to the industry efforts that preceded it.

1.2.2 The Development of Risk-Based Approaches to Flammables Management

The most recent evolution of flammables management is the use of risk-based approaches. In a risk-based approach, the expected frequency of a fire or explosion is quantified and combined with the predicted outcome of the fire/explosion to determine the risk of a potential hazard. To some extent, this evolution has been driven by the increasing availability of the computing power required to perform detailed analyses for thousands of scenario combinations that can be present in a modern process industry facility. This was also coincident with a rise of risk-based "culture" and risk-based regulations in Europe in particular.

The development of the quantified risk-based methodologies in recent years has been accompanied by tremendous advancements in the theory, tools, and software available to predict the consequences of fires and explosions. Although the methods for consequence analysis continue to improve, one can argue that the methods for consequence analysis are fairly mature and thus address half of the "risk equation":

$$\text{Risk} = f(\text{Consequence, Frequency})$$

or, in terms familiar to practitioners of layers of protection analysis:

$$\text{Risk} = \text{Consequence} \times \text{Frequency/Risk Reduction Factors}$$

The frequency side of the risk equation seems simpler conceptually and does not need to invoke Gaussian plume or computational fluid dynamics or other relatively higher mathematical solutions. In spite of this, or possibly because of this, the frequency of events has been a relatively neglected science. Now that is changing; because some regulators (mainly outside North America) require companies to perform quantified risk assessments, the regulators themselves have started to undertake standardization of frequency inputs to such studies. For example, some risk analysts are required to use specific values for the frequency of a leak of size X from a pressure vessel. While there is broad consensus on the values of many of these numbers in a "generic" situation, some inputs such as ignition probabilities are very situation-specific and so should be handled with greater rigor in many situations than is generally practiced.

Improvements to previous frequency/risk calculation methods are also timely given that the American Petroleum Institute (API) Recommended Practice 752 on building siting (API, 2009) permits use of risk as a basis for making building and personnel location decisions. Since the risk calculation for flammable events invariably incorporates a probability of ignition, greater precision and consistency in estimating this value are needed to ensure that risk assessments are both technically accurate and performed consistently across industry. Among other purposes, this book is therefore intended to provide new tools for users to comply with this API recommended practice and can be considered as a companion document to the CCPS book Guidelines for Evaluating Process Plant Buidings for External Explosions, Fires, and Toxic Releases (CCPS, 2012) as well as a supplemental resource for the CCPS book Guidelines for Enabling Conditions and Conditional Modifiers in Layer of Protection Analysis (CCPS, 2013).

1.2.3   Difficulties in Developing Ignition Probability Prediction Methods

From a mathematical point of view, determining ignition probabilities would seem to be a straightforward problem to solve—simply collect information or perform tests on events where flammables have been released and document the instances in which an ignition took place. However, the execution of this strategy is problematic from multiple perspectives, discussed next.

1.2.3.1 Data Bias

The simplest form of data analysis to develop ignition probability predictions is the following:

Probability of Ignition = Observed Ignitions/Observed Flammable Releases

There are numerous cases in which an event that resulted in a major fire or explosion has been documented in some form or another, especially in modern times. Thus there is some room for optimism in believing that the numerator for the equation above can be quantified with some level of confidence.

The denominator is another story. Ideally, releases that did *not* result in a fire or explosion should be documented with the same rigor as those that did. However, there is a much greater chance that a release that did not ignite will not be documented in an ignition database. It may be documented in other contexts, for example, for environmental reporting requirements. But it is much less likely that this data point will be delivered to someone developing an ignition probability database. Therefore, there could well be a bias toward concluding that ignition probabilities are greater than they actually are.

1.2.3.2 Experimental Problems

In the case of consequence model development, the industry has (at some considerable expense) conducted field test releases, fires, and explosions and measured the outcomes. Thus there are documented experiments that define the basis and calibration for the better consequence models that are available today.

In contrast, it is difficult to conduct experiments in ignition probability under controlled, real-life conditions. For example, no plant management could very well allow experimenters to perform dozens or hundreds of releases of a flammable mass into their unit to collect information on how many times an explosion resulted. A laboratory environment may be suitable for determining ignition probabilities for releases as they encounter a specific ignition source but can hardly be expected to replicate the hundreds of potential ignition sources available in an operating process plant.

1.2.3.3 Expert Opinion

There is a balance between what can be developed deterministically and the cost of doing so. The difficulty in obtaining objective ignition probability data has led many experts to propose values based on their personal experience. Such information can be valuable. However, it tends to suffer from two opposite problems: (a) widely different experiential outcomes dependent on context that is local to the individual observer and (b) replication of opinions in the literature, so that what appears to be a number of sources of the same probability value may in fact originate from some single source that may be lost in the sands of time.

For these reasons, the science of ignition probability estimation is not as developed as other risk input methodologies. As a result, risk analysts are often compelled to use ignition probability values that are very broad in nature (e.g., "immediate ignition probability of a light hydrocarbon" ~ 10%), whereas it is possible to envision one scenario fitting this description where the ignition probability is virtually nil (weeping flange leak from a remotely located butane bullet) and another scenario fitting this description where the ignition is almost certain to happen (release of a heavy hydrocarbon from a hydrotreater that is operating at a temperature well above the autoignition temperature).

1.2.3.4 "Conservatism"

Risk analysts generally try to use inputs that err on the side of conservatism to compensate for potential unknowns and uncertainties. However, "conservatism" is a difficult concept to apply to ignition probabilities, since choosing a "conservatively" high value for immediate ignition may lower the probability of a delayed ignition whose consequences could be much worse. Similarly, high ignition probabilities may preclude toxic outcomes that are more severe. This important concept is discussed in further detail in Section 4.1.2.

1.2.4 Missing Variables

An additional shortcoming of currently available ignition probability data is that there are variables that are known or suspected to be important but which are not easily quantified. Examples include the impact of the rate of ventilation in indoor releases or the effect of different electrical classification types in the area into which the material is released.

1.2.5 Summary of Industry Needs and Path Forward

The net effect of these and other shortcomings is not only a lack of information but also somewhat inaccurate risk analyses. It also means that risk managers have a greatly reduced set of risk mitigation options in their toolkit. For example, a risk manager may feel strongly that prohibiting vehicular traffic near a unit will reduce the risk. However, in the absence of quantitative justification of this conjecture, the risk manager will be unable to justify spending the money to implement the change. Thus it can be imagined that prudent risk reduction measures would not be undertaken without a recognized basis for doing so.

For this reason, the effort documented in this book has sought to expand upon both the accuracy and range of variables that can be considered in making an ignition probability estimate. The presumed accuracy of the methods in this book is discussed in Section 4.1.1.

There have been some notable efforts in recent years in developing ignition probability prediction methods. This book provides the reader with additional information on variables that have not been treated adequately up to this point and additional insights to many that have. As noted in the previous paragraph, this is not simply an issue of developing a more accurate tool to quantify risks in absolute terms; it is also critical to provide as many tools to the risk manager as possible to facilitate *reducing* the risks associated with flammable releases.

1.2.6   Applications for This Book

Estimates of ignition probability are useful in situations ranging from semiquantitative methods to conducting fully quantified risk analyses. The former typically require only crude levels of accuracy but are undertaken in situations where minimal effort can be justified to obtain the number. The latter requires a higher degree of accuracy, and the analyst should be willing to expend a little more effort to achieve that accuracy. This book is designed to satisfy a wide spectrum of users, allowing each user to expend the amount of effort necessary to achieve the level of accuracy desired.

While the potential applications of the tools in this book are considerable, the expected uses include the following (in increasing order of sophistication):

Process Hazards Analysis (PHA)—In PHA studies, teams frequently are asked to develop low-resolution measures of the risk associated with a particular scenario. For flammable releases, implicit in the estimate of event frequency is the probability that a release will result in a fire or explosion. The risk matrices commonly used for risk ranking in PHAs are usually delineated by order-of-magnitude levels of risk, and so the accuracy needed is minimal. The "Level 1" analysis as described in this book addresses this kind of application.

Layers of Protection Analysis (LOPA)—In LOPA studies, the probability of ignition is frequently used as a "conditional modifier" in assessing the frequency of a particular event outcome. The level of accuracy desired is greater than that needed for a PHA but only needs to be commensurate with the level of accuracy of the other inputs to the analysis. A "Level 1" or "Level 2" analysis as described in this book is appropriate for this application.

"Screening-Level" Quantitative Risk Assessment (QRA)—In QRAs, a yet higher degree of accuracy is desired, in order to be consistent with the level of effort and accuracy associated with the consequence side of the risk calculation. Other applications that are highly quantitative (e.g., determining optimal gas detector layout) will also benefit from a higher level of ignition probability accuracy. A "Level 2" or "Level 3" analysis as described in this book is justified for these applications.

"Detailed" Quantitative Risk Assessment—Some QRAs offer features that raise issues beyond those normally encountered in such studies. These include cases such as indoor flammable releases, where the ventilation rate or other factors not often described in existing literature are encountered. In these cases, a "Level 3" review as described in this book may be necessary.

1.2.7 Limitations in Applying the Approaches in This Book

This book provides useful algorithms for the purposes of predicting ignition probabilities in "typical" situations. However, in some atypical cases, the knowledge of the local plant staff in predicting ignition probability will be superior to that of any algorithms that could be developed. These include situations where the chemicals being released are pyrophoric or otherwise will react in the environment. Britton (1990a) describes the many nuances associated with the potential ignition of silanes and chlorosilanes, for example. Other situations include cases where the process operates under ultrasevere conditions, the chemicals decompose upon release, or the situation involves unusual chemicals or intermediates whose behavior is known only to company staff.

The methods described in this book should generally be applicable to normal plant layouts and operating conditions, for common chemicals such as hydrogen, hydrocarbons, and others that are not self-reactive or air-reactive. However, this book is not capable of anticipating all the possible physical and chemical influences that could be relevant to ignition of released masses.

Software users should have detailed knowledge of the areas where the software is being used for modeling. This will enable the user to account for the actual ignition sources and electrical classifications in these areas. This may require users to visit the areas where the releases are being modeled and to check out electrical classifications and ignition sources. The correct inputs are required to obtain the useful results from the software.

The algorithms may also be limited in accuracy in situations where the layout or operation of equipment is different than the typical range of operations in an onshore process plant. Thus users might find shortcomings in these tools as they are applied to offshore facilities or to releases during transport by ship, truck, or rail, for example.

It should also be noted that this tool is intended to predict the probability of ignition *given the precondition that a flammable cloud can reach an ignition source.* There are several reasons a flammable vapor cloud may not contact an ignition source, for example:

- The release rate is very small.
- The release is of a low-volatility liquid.
- The release is in a remote location.
- Drainage is present that directs a liquid spill away from ignition sources.

In cases such as these, the results of the models in this book are expected to have a higher degree of uncertainty, and the results should be used in conjunction with an event tree, augmented by a dispersion model and/or field walkdown, which describes the potential for ignition source contact.

Therefore, the tools described here should be used thoughtfully and need not be used for all possible circumstances. It is presumed that the user of the book is experienced enough in process plant operations to recognize situations in which the tool may, or does, give results that clearly do not reflect common experience for their chemicals, processes, and specific circumstances.

The product of this book is a set of algorithms that have been checked against the available literature and the experiences of people representing hundreds of years of plant operating experience in the aggregate. It is recognized that other models exist and in some cases coded into QRA and other software packages, and the developers of those models may consider their methods to be superior for the intended applications (which may be different than the scope of this book). Where this book provides enhancements to existing models, it may be possible to incorporate elements of this book into other, existing, models without using the entirety of the book. This should be done with appropriate care and testing of the result, since the methods in this book have been tested as a complete package and for a presumed experienced user.

1.3   IGNITION PROBABILITY OVERVIEW

1.3.1   Theoretical Basis for Ignition

1.3.1.1   What Is "Ignition"?

A logical first step in developing the basis for ignition probability is to define "ignition." The International Organization for Standardization (ISO, 2008) defines ignition as "initiation of combustion." Babrauskas (2003) notes that "combustion" is often not defined at all in texts dealing with it; he proposes a very simple description of combustion as "a self-sustained, high-temperature oxidation reaction." The NFPA (2012) defines it as "a chemical process of oxidation that occurs at a rate fast enough to produce heat and usually light in the form of either a glow or a flame." For the expected applications of the users of this book, "ignition" is defined as "the sudden transition to a self-sustained, high-temperature oxidation reaction."

This simple definition is suitable for the purposes of this book and eliminates events such as low temperature/rate oxidation reactions (e.g., rusting) or transient oxidation reactions [e.g., oxidation above the upper flammable limit (UFL), oxidation of "dry" CO] that are not of interest to us because they are not likely to be associated with severe-consequence scenarios.

1.3.1.2   How Does Ignition Occur?

Ignition can occur as the result of "autoignition" or "forced" ignition (Babrauskas, 2003), and in some cases, materials can self-ignite. As mentioned in another CCPS book (1993; page 318):

Apart from obvious ignition sources such as flames, several disparate groups of sources can be considered. These are:

- *Moderate temperature sources that may give rise to spontaneous ignition.*

- *Electrical sources such as powered equipment, electrostatic accumulation, stray currents, radiofrequency pick-up, and lightning.*

- *Physical sources such as compression energy, heat of adsorption, friction, and impact.*

- *Chemical sources such as catalytic materials, pyrophoric materials, and unstable species formed in the system.*

Ignition sources are often considered only in the context of the "fire triangle," whose sides comprise a fuel, an oxidant, and an ignition source (the three essential ingredients for most fires). However it is important to recognize that some materials can be "ignited" in the absence of an oxidant. Examples include acetylene and ethylene oxide (decomposition flames), and some metal dusts (reaction with nitrogen). Also, under process conditions, some materials may be "ignited" in the absence of oxidant even though at ambient conditions they may have a significant limiting oxidant concentration (LOC). An example is ethylene at elevated temperature and pressure (Britton et al. 1986).

Figure 1.1 illustrates an autoignition temperature-pressure diagram. For the purposes of this book, only the portion at atmospheric pressure is relevant, and only parts of this since the users of this book are assumed not to be interested in "cool" flames. Autoignition aside, 'point' ignition via a spark or other energetic source results in a reaction in which the local rate of energy release exceeds the rate of heat losses, with the result that the flame can propagate. The time scale for spark heating is on the order of 1 microsecond, with temperatures on the order of 55,000 °F (30,000 °C). This should be considered separately from "bulk" ignition such as autoignition, where ignitable mixtures are heated slowly.

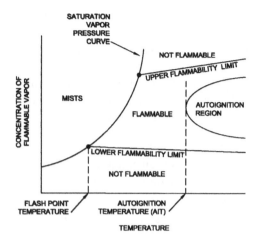

Figure 1.1. Relationships between various flammability properties (Crowl, 2003; adapted from Kuchta, 1985).

It must be recognized that many of the "ignition" examples are beyond the scope of this book to address quantitatively, and the user should not use this book to justify, for example, the use of a low ignition probability for the release of a pyrophoric material. Of primary interest in the scope of work of this book are autoignition and forced ignition, described in further detail next. In autoignition, the flammable mixture is present at a temperature hot enough to initiate and sustain the oxidation reaction. Forced ignition, on the other hand, requires a separate "agent" that supplies heat or energy.

Autoignition—Autoignition "is the result of self-ignition from any initial condition (temperature, pressure, volume) at which the rate of heat gain exceeds the rate of heat loss from the reacting system" (CCPS, 1993). It is difficult to define in real-life terms, since a flammable mixture simply does not transition from "no reaction" to "complete reaction" by increasing its temperature by some trivial amount. It is also a property that is difficult to predict using other readily available physical property measures such as boiling point, as illustrated in Figure 1.2.

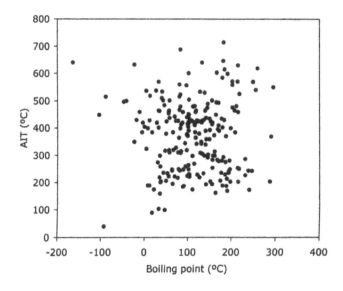

**Figure 1.2. Nonrelationship between AIT and boiling point
(Babrauskas, 2003).**

Autoignition temperature (AIT) is a function of the oxidation reaction kinetics and can be related to some extent to the molecular structure of the chemical (such as the degree to which there is branching in the molecule). Babrauskas (2003) notes an example of the branching of two octane isomers—normal octane and isooctane. The former has a relatively low AIT of 220 °C and the latter has an AIT of 415 °C. This dependence of the AIT on the degree of molecular branching has been described by Zabetakis (1965) for some classes of compounds and is illustrated in Figure 1.3. Although dated, this seminal reference is still readily available and covers a broad range of flammability issues.

The effect of chain link on AIT has been ascribed to the influence of chain length on the frequency of chain breaking into radicals. This can also be related to the "cetane" and "octane" numbers used to describe the performance of diesel and gasoline engine fuels, respectively. Kinetic theory says that increased temperature increases the probability that a molecule will be provided with sufficient energy to break apart and subsequently take part in radical reactions with oxygen.

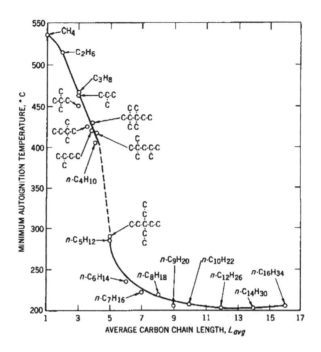

Figure 1.3. AIT as a function of chain length (Zabetakis, 1965).

Other Self-Induced Ignition—A likely source of immediate ignition for released flammables is static discharge. In this case, the movement of the process fluid through a pipe or other equipment prior to discharge may be greater than the normal flow, thus generating electrical charges that are greater than were anticipated by the designers. Upon release, this charge may be imparted to the released material.

Britton and Smith (2012) provide a detailed methodology for assessing charging currents caused by flow of low-conductivity liquids. Appendix A is helpful for identifying "static accumulating" liquids. Methods are presented for assessing the probability of ignition. Flash point safety factors are given and it is shown how the decreased ambient pressures associated with high elevations (such as in mining operations) can be accounted for, both for single-component liquids and complex mixtures such as diesel oil.

Forced Ignition—Ignition of a flammable mixture can obviously occur if there is a preexisting flame present. Note that all flames are not sufficient to the task. For ignition to occur, the flame must:

- have a sufficient temperature
- be of sufficient size and shape
- be applied for a sufficient time

The required flame application time may be on the order of milliseconds but may also be substantially longer.

If a sufficient flame is not present, the ignition must be "forced" in some other manner. The simplest forced ignition is supplied by a hot surface. An electrical discharge, while hot, also provides excited chemical species that can aid in propagating the combustion reaction. A spark can be provided by supplying a sufficient voltage or by opening or closing an electrical contact—for example, if an electrical supply wire is broken through mechanical damage.

The most effective spark ignition sources are when the sparks are delivered quickly (microseconds) and over a limited space ("point source"). Little energy may be needed; the energy required for ignition is often under 1 millijoule (mJ). However, applications of larger energies may not be sufficient if the energy input is too spread out over time. In process plant releases, the time issue is generally not of concern with respect to the energy source, since the energy sources tend to be sustained for some period during an emergency event. Time may still be a relative variable, however, but more with respect to the duration of a release and its ability to find an ignition source in the time available.

1.3.2 Key Ignition Factors Related to the Properties of the Fuel and Available Surrogates That Can Be Used for Developing Probability of Ignition Predictions

A number of chemical properties are significant in determining the propensity of a release to ignite. These properties have been determined experimentally for many chemicals, but not all. For the purposes of developing a predictive tool, the following text discusses those factors that are relevant as well as the availability of data for a broad spectrum of chemicals. It should be noted that while the factors that follow are relevant, some are not included in the subsequent model development, generally because the variable is closely related to one or more other variables that are being used and so does not need to be repeated.

Where data is not readily available for some chemical properties, more widely available chemical properties are proposed as surrogates. The CHETAH program of ASTM International (ASTM, 2011) calculates various flammability parameters for hydrocarbons (C+H) and molecules containing C+H+O atoms (alcohols, ethers, ketones, etc.). The same methodology is applied to organic C+H compounds that contain nitrogen atoms (amines, etc.) but with less accuracy.

Estimates are allowed for lower flammable limits of some chlorinated organic species.

Mixtures of chemicals introduce another level of complexity. The treatment of mixtures is discussed in Section 2.10.2.

1.3.2.1   Flammability Limits

Most fuels will burn only within a given range of concentrations in air. If the fuel mixture is too "rich," there is not enough oxygen available to initiate combustion. If the fuel mixture is too "lean," not enough fuel is available. The importance of the flammability limits to the ignition probability question lies in the size of the cloud that is in the flammable range. If the flammability range is very broad, the resulting dimensions of the flammable cloud are also very large (given all other things being equal).

Note that the flammability limit varies with the temperature of the chemical, becoming broader at higher temperatures. However, for most chemicals the effect of flammability limits is not as great as for some other chemical properties, and hence the impact of temperature on this flammability limit is small enough that it can be neglected, inasmuch as temperature is incorporated directly as an input for other reasons. It is also true that, in contrast to a laboratory flammability test, a release into the open environment will present a range of concentrations such that if the process stream is flammable at all, it will pass through the flammability range during its dispersion.

Lastly, many of the applications for this tool such as QRA will also involve use of dispersion models. It must be remembered that such models may have something like a factor of two accuracy, and any given dispersion may be nonuniform—with pockets of flammability beyond the predicted lower flammable limit (LFL), or places of nonflammability within the predicted LFL distance. For this reason, many models have outputs that report ½ LFL distances as well as LFL. It is not within the scope of this book to discuss whether the ignition probability models here should be applied to LFL or ½ LFL or some other endpoint. This is a decision left to the user, knowing the larger context and conservatisms built into the other methods used in the combined analysis.

1.3.2.2   Flash Point and Flame/Fire Point

The more widely used of these two terms, the "flash point," is defined as the lowest temperature at which a liquid produces a flammable vapor (has a vapor pressure that produces a vapor concentration equal to its LFL concentration). The flame (fire) point may be a better measure of the propensity to ignite, however, as it is defined as the lowest temperature at which a liquid creates a *continuous* flame. In any case, it must be noted that neither flash point ("open cup" or "closed cup") nor flame/fire point measurements are perfect reflections of behavior in an open-air release; for our purposes they are merely surrogates that are intended to describe a relative tendency to ignite.

The Society of Fire Protection Engineers provides some overview and background data, although not probability data, in their "Handbook of Fire Protection Engineering" (SFPE, 2008). Specifically, physical property data such as flammability ranges, autoignition temperatures, flash points, and more are included. In addition, there is a discussion of the basis for ignition of liquids, and the relationship between flash point and fire point, illustrated in Table 1.1.

The physical basis for differences in these values relates to the manner in which the tests are conducted. The closed-cup flash point apparatus allows vapor to accumulate above the liquid until it reaches equilibrium, whereas the open-cup apparatus allows the vapor to diffuse away. Thus the latter produces higher ignition temperatures. Higher temperatures still are required to reach the "fire point," which consists of a self-sustaining diffusion flame. The fuel-rich fire point flame can be contrasted with the lean flames developed for the flash point tests; in the flash point tests the flammable vapors are completely consumed and not sustained.

The flame/fire point is not available for many chemicals, however. Fortunately, although there are significant exceptions, the flame/fire point is typically 5–15 °F higher than the flash point (closed cup). Therefore the flash point is chosen as a reasonable surrogate to the flame/fire point. In turn, several investigators over the years have explored the relationship between flash point and boiling point. A good synopsis of this work is described in Catoire and Naudet (2004), which points toward an example of a correlation that is accurate enough for the purposes of this book that boiling point might be used as a surrogate variable in place of flash/fire point. Alternatively, the flash point might be predicted using methods such as those described in Rowley et al. (2010).

That said, there are difficulties in using boiling point—for example, when mixtures are involved. In these cases it is typical practice to pick the temperature at which the most volatile 10% of material has flashed, although this is imperfect.

Table 1.1. Some values of closed-cup flash point,
open-cup flash point, and fire point temperatures (SFPE, 2008)

	Closed-Cup FP	Open-Cup FP	Fire Point
n-Hexane	-22 °C [-8 °F]	N/A	N/A
n-Heptane	-4 °C [25 °F]	-1 °C [30 °F]	2 °C [36 °F]
Methanol[b]	12 °C [54 °F]	1.0, 13.5[a] (°C) [34, 56[a] °F]	1.0, 13.5[a] (°C) [34, 56[a] °F]
p-Xylene	25 °C [77 °F]	31 °C [88 °F]	44 °C [111 °F]
n-Butanol	29 °C [84 °F]	36 °C [97 °F]	36, 38, 50 (°C) [97, 100, 122 °F]
n-Nonane	31 °C [88 °F]	37 °C [99 °F]	42 °C [108 °F]
JP-6	N/A	38 °C [100 °F]	43 °C [109 °F]
n-Dodecane	74 °C [165 °F]	N/A	103 °C [217 °F]
Fuel oil no. 2	124 °C [255 °F]	N/A	129 °C [264 °F]
Glycerol	160 °C [320 °F]	176 °C [349 °F]	207 °C [405 °F]

N/A = Not available
[a]The lower value was obtained from ignition by a pilot flame; the higher number is from ignition by a spark.

1.3.2.3   Vapor Pressure

The vapor pressure of a material is a function of the chemical itself and the temperature at which it is released. This material property describes its propensity to vaporize if it is released in liquid form and thus is a key factor on how large a vapor cloud could become—which in turn relates to the odds that it will encounter an ignition source.

Since the vapor pressure is dependent on the chemical's temperature, using vapor pressure as an explicit input to an ignition model would require building vapor pressure/temperature curves into the model for the large number of chemicals potentially of interest. For this reason, the chemical's boiling point (a readily available property) relative to its release temperature (known by the analyst) is used as a surrogate for vapor pressure. The treatment of mixtures in this regard is discussed Chapter 2.

1.3.2.4 Autoignition Temperature (AIT)

Also known as the "ignition temperature" or the "spontaneous ignition temperature," the AIT is the temperature at which a material will ignite in the absence of an external ignition source. It presumably is the point at which a fire is guaranteed to happen and an explosion is guaranteed not to happen (since there is no time for an unignited cloud to accumulate).

However, the reality of AIT is not as clear as that. The measured AIT has a number of shortcomings related to predicting the actual spontaneous ignition of a chemical, including:

Test Apparatus Variability—The test apparatuses used for measuring AIT vary considerably, with container surface effects greatly affecting the results. Therefore the AIT values reported for individual chemicals can vary by 200 °F or more. Usually the effect is that a chemical has to be at a temperature that is considerably higher than the reported AIT in order to actually autoignite, since the medium into/onto which a material is released will cool the release. The API Risk-Based Inspection protocol (API, 2008) takes this into account by assuming that AIT has no effect on ignition probability until the release temperature is 80 °F higher than the reported AIT. However, it should be noted that, in principle, if a released cloud is *confined* in a hot space larger than the test apparatus used to measure AIT, the actual AIT could be lower than measured.

Surface Encountered—There are cases where autoignition has been reported at temperatures below the AIT. This is generally attributed to a hot liquid release encountering a high surface area space, such as might be present as insulation on a high-temperature vessel. This can also happen if the release reacts with the surrounding surface (e.g., rusty steel). These situations are quite dissimilar to an AIT test apparatus, which is typically a clean, smooth surface.

The AIT is clearly an important factor in the ignitability of a flammable release. However, since it is an imperfect measure in the real world of a release, it is not treated as a discrete value above which ignition is 100% certain and below which autoignition is impossible. Rather, it is assumed that there is a range of temperatures above and below the reported AIT at which autoignition can actually occur. Appendix A provides AIT values for some common industrial chemicals.

1.3.2.5 Minimum Ignition Energy (MIE)

The MIE is the most significant chemical property in the forced ignition of a flammable mixture. As with AIT, the MIE is not readily relatable to commonly available physical properties. In fact, about the only other parameter to which it is reliably related is the "quenching distance," which is itself a comparatively arcane measure of the "minimum dimension that a flame kernel must acquire in order to establish a freely propagating flame" (Babrauskas, 2003) in order to avoid self-extinguishment due to having greater heat losses than heat gain.

Probably the most useful relationship between MIE and another chemical property is that by Britton (2002), who correlated MIE to the heat of combustion per mole of oxygen consumed. This correlation is explored further in Chapter 2. MIE may also be related to more fundamental chemical properties such as the Lewis number (ratio of thermal diffusivity to mass diffusivity) and activation energy (Tromans and Furzeland, 1986).

There is a broad range of MIE values among commonly used chemicals (having typical MIEs > 0.2 mJ), with hydrogen (MIE ~ 0.017 mJ) being on the extreme low end. However, such disparities in MIE may not be as significant as they appear at first glance. Dryer et al. (2007) note that the MIE for hydrogen (like other materials) occurs near the stoichiometric concentration—which is 29 volume % in air. However, at the LFL, the MIE "is more similar to that of methane" (Dryer et al., 2007). Appendix A provides some reported MIE values that range almost six orders of magnitude.

1.3.2.6   Multiple Release Phases

The phase of the material being released is relevant for two primary reasons. Most obvious is the need for the discharge to be a vapor, form a vapor, or be sufficiently atomized to have access to the oxygen necessary for ignition.

However, the phase is also important with respect to electrical charge buildup. The Occupational Safety and Health Service (OSHS, 1999) notes that "pure gases discharged at high velocity through jets under conditions where neither liquid droplets nor solid particles are present, seldom acquire sufficient static charge to result in ignition. However, when the gases contain liquid droplets or solid particles, or when these are formed during the discharge, sufficient charges can accumulate to ignite flammable vapours present."

Thus streams that form droplets during discharge, such as liquefied petroleum gas (LPG), or releases that are accompanied by particulates discharged from the same equipment are more prone to static ignition than they might otherwise be based on their chemical properties alone. Static development on droplets or particulates can also explain anecdotes such as ignitions caused by application of water sprays and carbon dioxide intended to blanket flammable releases. Static formation on particulates may also help explain the highly disparate anecdotal evidence for either high or low probabilities of ignition of hydrogen.

1.3.2.7   Summary of Chemical Property Factors

In order to minimize the effort for the user, it is useful to reduce the number of variables that are input as much as practical without greatly affecting the accuracy of the results. For this reason the flash/fire point and vapor pressure factors discussed above will be "bundled" into a single variable that is dependent on the normal boiling point or flash point of the material. The flammable range of a material is also assumed to be indirectly bundled into the boiling/flash point factor according to arguments made in Chapter 2. However, AIT and MIE are not reducible to more common physical measures and so will be utilized as is.

1.3.3   Key Ignition Factors Related to the Release Source

Many of the variables that follow depend to some extent on how the release event unfolds and the physical layout near the point of release. The major relevant factors are discussed next.

1.3.3.1   Release Rate

Previous incident data indicate that the larger the release rate, the greater the probability of both ignition overall and explosive ignition. Presumably this is related at least in part to the simple fact that the greater the release rate, the greater the size of the flammable cloud that results, and hence greater opportunity to reach ignition sources. A greater release rate may also be associated with greater static formation, discussed later.

1.3.3.2   Release Pressure

The release pressure is obviously related to the release velocity, finer aerosolization of liquids, and the potential for static discharge near the source. These features are presumed to represent the key influences of pressure on ignition probability, at least for liquid or two-phase releases, with higher pressures leading to higher chances of ignition.

The situation for vapor releases may be more complex. There are anecdotal indications that higher pressures lead to higher chances of ignition; the ignition mechanism in this case may result from electrical buildup on particles (e.g., scale) that are discharged with the vapor. However, there is a potential countereffect of high pressure, namely, the potential for an incipient flame to be "blown off" the end of the flammable cloud. In principle, this could occur when the flame speed of the flammable gas is less than the velocity of the flammable gas jet leaving the release source. In practice this phenomenon has been observed by Swain et al. (2007) and others and is discussed further in Chapter 3.

Another possible mechanism for ignition at low pressures is suggested by Britton (1990a). This mechanism presumes that pyrophoric solids are present at the point of discharge but are not exposed to air unless the release velocity is very low, typically, at the end of a release event. At the low pressures, air entrainment (dilution) is reduced, and the pyrophorics may be exposed to air that migrates into the equipment that is the source of the release.

1.3.3.3   Release Temperature

Aside from the obvious importance of release temperature relative to the AIT, the temperature has other influences on ignition potential; among them:

- Increased temperature often broadens the range in which a cloud is flammable or otherwise lowers the threshold for ignition to take place.

- The temperature influences the degree to which a liquid or two-phase release will vaporize and to a lesser degree affects the buoyancy of a gas cloud.

1.3.3.4   Event Duration

For a given release rate, a longer duration event may be more likely to ignite than a short duration event because: (a) there will be greater opportunity for the ignition sources to act upon the vapor cloud and (b) the vapor cloud itself may be larger and thus cover more ignition sources until it reaches its maximum profile. There is probably a time limit beyond which an ignitable cloud will not ever ignite—that is, given the conditions for ignition that last for 10 minutes or so without ignition, the event most likely will never ignite for reasons that are outside the ability of this book to quantify.

1.3.3.5   Static Discharge At/Near the Point of Release

The environment at or near the point of release is critical, primarily in determining whether immediate ignition occurs or not. The configuration of the equipment feeding the release and the geometry of the release hole are important (along with the fuel ignition properties) in determining whether a static discharge occurs that ignites the release at the source. Perhaps indistinguishable in impact, but also important, is whether any static or other ignition sources are present in the immediate vicinity of the release. The issue of static discharge is an important one that is discussed in further detail below.

Static Ignition—Two excellent books on the subject of static ignition hazards have been published by CCPS (Britton, 1999; Pratt, 2000). A synopsis of these, as they apply to the ignition of released flammable masses, follows.

Britton provides a summary of typical ignition sources and energies, and their applicability to a variety of flammable masses (Figure 1.4).

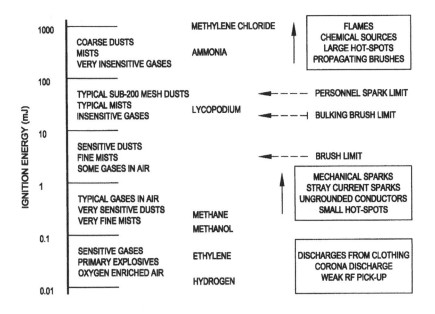

Figure 1.4. Ignition energies of various materials and types of ignition source that may ignite them (updated from Britton, 1999).

Most of the materials of interest to the readers of this book have ignition energies under 1 mJ, and the focus of this book is therefore on those situations. Although other ignition sources exist (e.g., lightning), they are generally not relevant to this work because they are transient events that have a very small probability of occurring coincident with a flammable material release.

The differences between the various types of static discharge are very briefly discussed next, largely based on illustrations in Pratt and Britton. The reader is invited to review these books for a much more thorough understanding of this subject.

Corona Discharge—Figure 1.5 and Figure 1.6 depict a corona discharge. Corona discharge occurs when a sharp point is present in the air near a charged surface, inducing ionization in the air. By definition, the charges are diffuse, and so this ignition source is capable of igniting only the most sensitive chemicals.

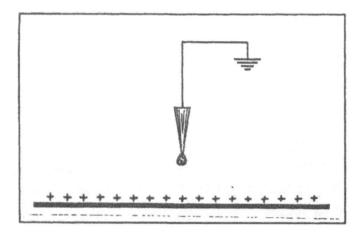

Figure 1.5. Corona discharge (Pratt, 2000).

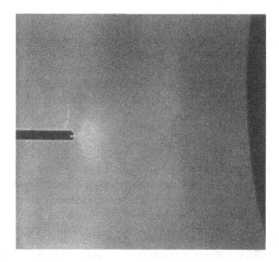

Figure 1.6. Corona discharge in point-sphere gap (Britton, 1999).

Brush Discharge—Brush discharge is illustrated in Figure 1.7 and Figure 1.8.

In contrast to corona discharge, the electrode in a brush discharge has a shape that is curved in some way (that is, not sharp). Tools, extensions from vessels, and fingers are examples. As before, the ignition potential is created with a charged surface (e.g., pipes, mists), but this is much more energetic than a corona discharge.

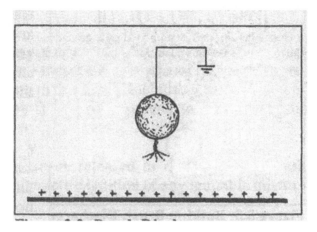

Figure 1.7. Brush discharge (Pratt, 2000).

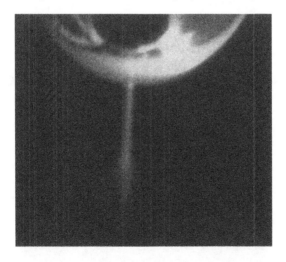

Figure 1.8. Positive brush discharge from negatively charged plastic to grounded sphere (Britton, 1999).

Bulking Brush Discharge—This type of discharge is associated with an accumulation of powder and so will usually not be relevant to the intended audience of this work. However, it could be quite relevant to those who handle solids that off-gas flammable vapors while in storage (noting that storage applications are outside the scope of this book). Britton and Smith (2012) also show that similar discharges can be produced in liquid systems.

Propagating Brush Discharge—This extremely energetic form of brush discharge is depicted in Figure 1.9 and Figure 1.10. The surface must have an extremely high charge density that is supported by a grounded conductor for this type of discharge to occur, since this condition allows most of the electrical field to exist between the surface and the backing rather than in the air where it would dissipate.

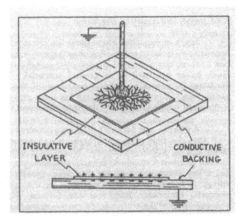

Figure 1.9. Propagating brush discharge (Pratt, 2000).

Figure 1.10. Propagating brush discharge on charged layer initiated by grounded electrode (Britton, 1999).

Spark or Capacitor Discharge—This type of discharge is of primary interest in the ignition of flammables in a process environment. The spark discharge occurs between capacitors. The discharge is shown in Figure 1.11 and Figure 1.12.

Common objects in a process plant can store a static charge and thus be a sufficient capacitor for the purposes of providing an ignition source: buckets, tanks, vehicles, machinery components, and humans are just a few of the possibilities. Some chemicals (e.g., low-sulfur diesel fuel) have recently been discovered to harbor charges.

Release Equipment Type—There is some evidence that the type of equipment may influence its propensity to build a static charge, all else being equal. For example, it is noted in Glor (1999) and others that static buildup in piping systems is a function of the velocity of the fluid through the pipe. This may be relevant for pipe rupture cases discussed in this book in which the velocity through the pipe may be much larger than is the case in typical process situations (and for which static dissipation measures have presumably been designed).

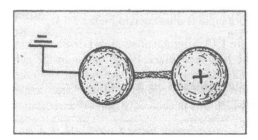

Figure 1.11. Spark discharge (Pratt, 2000).

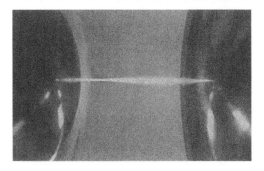

Figure 1.12. Spark discharge between spherical electrodes (Britton, 1999).

Summary—There are potentially several types of static discharge that could ignite a flammable release near the source (or farther away). For the purposes of predicting ignition probabilities, the distinction between the different types of static ignition is probably not important. Even if it was important, there is likely no practical way to account mathematically for the numbers and strengths of the sources in normal operation, let alone in the possible chaos of damaged equipment coincident with the release of material. For this reason, generalized approaches to predicting ignitions near the source (e.g., based on size of release, average equipment density) are probably as good as more complex schemes that might be devised.

1.3.3.6   Restrictions at Point of Release

Dryer et al. (2007) performed a series of experiments with hydrogen (and to a much lesser extent natural gas) in which the presence of obstructions or confinement at the point of release (in this case, tubing/fittings downstream of a hydrogen cylinder rupture disk) can influence the odds of ignition. He concluded that "similar phenomena apparently do not necessarily occur for sudden boundary failures of storage vessel or transmission piping into open air that have no downstream obstruction." The same is considered credible for methane and natural gas, but not for heavier hydrocarbons.

The presumed principle is illustrated in Figure 1.13.

As seen in Figure 1.13, after the pressure is released from the rupture disk, the high pressure of the hydrogen shock heats the air in the confined space of the discharge piping. The effect is enhanced when there are obstructions in the line such as fittings that promote mixing. There are ranges of pressure and configuration over which this phenomenon appeared to occur, discussed further in Chapter 3.

The ignition phenomenon was only observed for release pressures greater than about 200 psig. The requisite confinement was effective down to a discharge piping length of about 1.5 inches. However, ignition did not occur at lengths greater than about 10 feet, which was attributed to combustion heat removal by the piping.

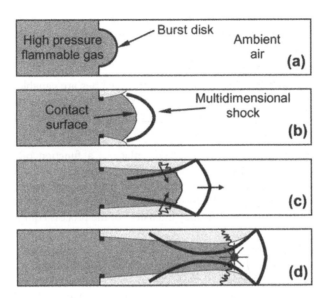

Figure 1.13. Schematic of shock wave formation downstream of a release of high-pressure gas into a confined space (Dryer et al., 2007).

1.3.4 Key Ignition Factors Related to the External Environment After the Release

1.3.4.1 Existing Flames

Flammable mixtures can be ignited by preexisting flames such as those from fired heaters. These may be logically treated as point sources of ignition, and accounted for individually. Alternatively, it may be simpler to assume a certain density of such sources in a normal process plant.

1.3.4.2 Burning Particles and Sparks

Flammable mixtures may be easily ignited by the energies available in sparks generated by rotating cutting equipment, from friction in machinery, or through impact of sparking tools or falling objects. There is no ready measure for this, and so it is treated as an "area" source of ignition—that is, in a given area of a process plant it is assumed that there may be some concentration of such ignition sources, but it is not practical to try to quantify them except perhaps through the electrical classification of the area.

In principle, sparks could be caused by someone dropping a tool while escaping from a flammable gas cloud. However, this does not appear to be a significant cause of fires or explosions, and the use of nonsparking tools instead of ferrous tools should not be considered a preventive measure with quantifiable benefit (API, 2004).

1.3.4.3   Electrical Equipment

Equipment including electronic circuits and power lines are capable of providing the energies required to ignite a flammable mixture. Objects close to ground level with overhead ultrahigh voltage (i.e., >800 kV) transmission lines are situated in an electric field which may be on the order of 5 kV/m (CCPS, 1993).

Whether or not ignition actually takes place is a function of the available voltage, current, and other factors. Because of the wide range of electrical sources that are available in a typical process plant, either they may be assigned an ignition "strength" based on the type of exposure that is present (for point sources) or an assumption can be made based on the "intensity" of the process equipment.

1.3.4.4   Hot Surfaces

Hot surfaces will be present in a typical process unit in some form (e.g., hot piping, motors). However, hot surfaces do not offer the same ignition properties as sparked ignition—that is, the temperature at which a surface will ignite a material is not the same as the laboratory autoignition temperature.

The reason is attributed to convection of heated gases away from the hot surface, which, if rapid enough, does not allow enough eduction time for the ignition to take place. In contrast, an autoignition apparatus uniformly heats and contains a mixture that cannot escape. Duarte et al. (1998) illustrated this phenomenon (Figure 1.14), described it in some detail, and conducted bench-scale experiments that supported this explanation.

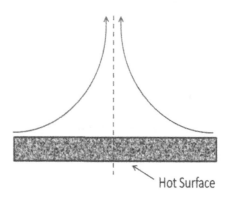

Figure 1.14. Flow patterns of a horizontal upward-facing heated surface (Duarte et al., 1998).

A hot surface cannot be treated as a point source in the same sense as a spark would be, but it is assumed that hot surfaces are limited in dimension. Figure 1.15 illustrates the effect of the area of the hot surface on the ignition temperature of various fuels; note that the largest area considered in the figure is still small enough to be considered as a "point source" for the purposes of these methods.

Figure 1.15 also illustrates that the ignition temperatures for hot surfaces can be substantially higher than those from an autoignition apparatus (e.g., AIT of hexane = 223 °C; of hydrogen = 528 °C; of diethyl ether = 195 °C). API has also compared experimental AIT values with ignition temperatures observed over hot surfaces (API, 2003).

On this basis, hot surfaces can be considered on either a point-source or area-source basis, but with assumed lesser "strength" than the preceding ignition sources. However, certain administrative practices may be considered to modify hot-surface ignition sources; for example, if vehicular traffic is prohibited from the vicinity of the release location. Note that later in this book vehicles are not treated specifically as hot-surface ignition sources, since the cause of ignition may be either a hot surface or the vehicle's ignition system.

Further discussion of this topic can be found in Sections 3.3.4 and 3.3.5.

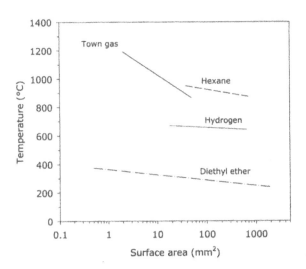

Figure 1.15. Effect of hot-surface area on ignition temperature (Babrauskas, 2003).

1.3.4.5   Discharge into Open or Closed Space?

The vast majority of ignition data has been collected for releases into open (outdoor) spaces. However, if an indoor release occurs, it may be more likely to ignite because the vapor cloud will be contained and will not disperse as readily without wind or turbulence to promote the dispersion—in fact, the dispersion will probably be dominated by the release of jet-induced turbulence. This might be considered the general rule; in some cases the confinement may result in the space being *un*ignited due to being above the upper flammability limit.

Dilution of the vapor cloud will occur as the result of natural or forced ventilation in the room. Whether or not the ventilation rate is sufficient to reduce the size of the cloud that is in the flammable range is a function of the size of the release, size of the room, ventilation rate, and release/ventilation/room geometries.

In terms of impact, a release inside a building also increases the probability that the ignition will result in an explosion, since the building interior provides a degree of confinement that, as explained in Section 1.4, can promote damaging blast wave overpressures.

1.3.4.6   Electrostatic Sources, Including Human Sources

The basis for electrostatic discharge was discussed earlier and doubtless has a role to play in far-field ignitions as well as "immediate" ignitions. The focus of this discussion is therefore on human sources of electrostatic discharges. This topic has been discussed by Johnson (1980) and others.

Johnson relates these requirements for human electrostatic ignition of flammables:

- Presence of a vapor/air mixture within a fairly narrow concentration range
- Sufficient electrostatic charge generated on the person
- Storage of the charge on the person for a significant length of time (which requires low absolute humidity)
- Large or grounded object available to discharge the spark

Johnson goes on to cite conflicting reports on the viability of humans as electrostatic ignition sources but through experimental studies concluded that the energy required for human electrostatic ignition of flammables was only 2–3 times that of a capacitive spark for ignition of acetone/air mixtures. Furthermore, it was anticipated that any vapor/air mixture with an MIE of 5 mJ or less could be ignited by a person. A value of 25 mJ being discharged from people due to static charges has been suggested (OSHS, 1999), although other values ranging from 10 to 30 mJ are also assumed.

1.4   CONTROL OF IGNITION SOURCES

1.4.1   Ignition Source Management

By far the most common method for managing ignition sources is through instituting proper hazardous area classifications as per various currently available fire protection standards. However, there are limits to the effectiveness of this and other ignition control measures. This is discussed next.

1.4.1.1   Hazardous Area Classification

All modern petrochemical facilities are designed to one or more industry standards that regulate the type of ignition sources that are allowed in a given area based on the expected flammable environments to which the ignition source will be exposed. Some commonly used English language standards are from the National Fire Protection Association (NFPA, 2012, and others), British Standards (BS, 2009, and others), and the American Petroleum Institute (API, 2002); examples of various standards as applied to a storage tank are shown in Figure 1.16.

Attributes of ignition control measures are also described by CCPS (2012).

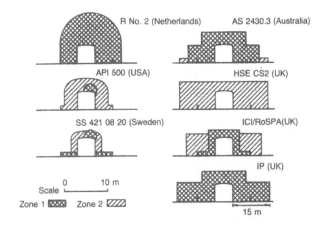

Figure 1.16. Depictions of hazardous area classification (Mannan, 2005).

In these standards, the environment at a particular location is characterized in broad terms of the frequency and persistence of flammable atmospheres. These standards have enjoyed great success over the years, although for the purposes of this book, they have two significant shortcomings:

- They are designed to address continuous or incidental (e.g., maintenance activity) releases but not major losses of containment.
- The integrity of the devices may be compromised by the same event that ultimately becomes ignited if the initiating event is energetic (e.g., dropped object).

Therefore, while area classification is an important component of an ignition prevention program, it is not fool-proof.

1.4.1.2   Work Permit Systems

It is assumed that any facility utilizing this book will have an active and effective work permit system that is designed to minimize ignition hazards from hot work such as welding, grinding, and the like. However, even a "best of class" work permit system does not provide perfect control—even if hot work is discontinued at the first signs of a flammable release, residual hot surfaces may still provide an ignition source. Job safety analyses provide a supplemental level of safeguard in cases where there is a hazard introduced by simultaneous but unrelated activities, such as hot work in the area where a pump is being opened for maintenance.

Daycock and Rew (2004) cite statistics regarding how the failure of work permit systems have led to ignitions that might otherwise have been avoided. Notably, around one-fourth of the fire/explosion events they reviewed were associated with permit-to-work failures: "*A common theme, which was apparent across all of the incidents reviewed, was that the hazards associated with the task were not adequately assessed pre-job.*"

The effectiveness of a work permit system in preventing ignition is important but difficult to quantify. The statistics mentioned above are likely not specifically attributable to the magnitude of release events of interest to the readers of this book. Therefore no attempt is made here to quantify the effect of work permit systems—good, bad, or otherwise, although this may be worthy of consideration by safety managers. However, the algorithms developed in this book do consider the broader issue of ignition source control, including and beyond that related to work permits.

1.4.1.3   Vehicle Limits

Some fraction of historical fire and explosion events were found to have ignited due to the presence of vehicles in the area of the release. There are at least two mechanisms by which this can happen: (1) the gas cloud is ingested into the combustion system of the vehicle or (2) the gas cloud is ignited by a hot surface on the vehicle such as the engine or the exhaust system.

For these reasons, it is appropriate to give credit for site programs that prevent vehicular traffic from being near to an operating unit—or to penalize locations that do allow such traffic except under controlled situations (e.g., hot-work permit).

1.4.1.4   Control of Electrical Ignition Sources

The algorithms in this book assume that typical electrical classification and work permit programs are in place for the facilities being evaluated. However, it should be possible, in principle, to provide credit for instituting controls that are more aggressive than those required under the usual standards, and this can be considered in developing ignition prediction algorithms.

Also important is control of frequent/continuous static sources such as accumulation of static in the handling of hydrocarbons, presence of electrical lines, or accumulation of static charge (particularly indoors) from clothing, footwear, or floors.

Proper grounding is the solution to each of these issues, although it is also appropriate to consider that even properly designed electrical equipment (e.g., those with proper location and protection of instrument/power cables, location of high voltage lines) have the potential to ignite a release.

In the case of indoor releases, Johnson (1980) and others have suggested the use of conductive (e.g., lead) floors, conductive footwear, and maintaining the relative humidity above 60% for rooms operating at typical temperatures.

Lightning is also an obvious ignition source and to some extent can be mitigated through lightning protection systems. For the purposes of this book it is assumed that the probability of lightning coincident with a major flammable release is remote, although there are certainly instances where lightning can lead to a flammable release.

1.4.2   Minimization of Release

There are several methods commonly employed to minimize the amount of material release or to contain it or otherwise reduce its potential for spreading or ignition. These include:

- Suppressing the release with water or foam
- Using fluid curtains or fogs to contain the spread of a release
- Secondary containment to minimize vaporization from a pool (e.g., dike)
- Leak detection/isolation systems

These are described in CCPS (1997), Fthenakis (1993), Murphy (2009), and others.

Note that there have been occasional reports of fire/explosion suppression systems actually *causing* an ignition due to static generation or other factors (Britton, 1999). For the purposes of this book it is assumed that such systems are properly designed and maintained for the application and at worst provide no benefit.

1.5   VAPOR CLOUD EXPLOSION PROBABILITY OVERVIEW

Flammable releases can have various outcomes (illustrated later in event tree form in Figure 2.1). Explosions are one of these possible outcomes. Vapor cloud explosions (VCEs) are caused by combustion of a dispersed cloud of vapor in a congested and/or confined volume, which is defined as a space containing turbulence-inducing obstacles. The VCE produces an overpressure wave that propagates into the surroundings. For the purposes of this book, explosions are assumed to occur only if ignition is delayed, such that a sufficiently large, unignited vapor cloud can form.

The variables that determine whether an explosion takes place or not include the following:

- The fundamental burning velocity of the fuel
- The degree of congestion in the space in which the flame front develops
- The degree of confinement in the space in which the flame front develops

The greater that each of these variables is, the greater is the propensity for an explosion. Note that the last two variables may well depend on the magnitude of the release—that is, the larger the release, the greater the chance that the resulting vapor cloud can reach an area or areas of sufficient congestion/confinement to generate an explosion.

The fundamental burning velocity is an intrinsic property of the fuel. The congestion and confinement are properties of the environment into which the fuel is released but can be described in general terms for the purposes of developing an ignition probability model. Alternatively, the analyst can use standard blast overpressure models described in CCPS (2010) to determine whether an explosion is likely.

1.5.1   Explosion Venting

An incipient flame front may not develop damaging overpressures if there is sufficient venting present. In the case of an outdoor release, when the ignited mass expands into the surrounding environment (assuming that it is not unduly restricted in some way), effective mitigation can be provided by sufficient spacing between equipment such that dangerous flame front velocities never develop.

Inside a building, mechanical venting may be needed to prevent or minimize an explosion/detonation. Since this discussion is limited to preventing explosions outside of process equipment, this can primarily be accomplished through use of blowoff panels. As noted in the SFPE Handbook (2008):

"The most effective explosion venting systems are those that deploy early in the deflagration, have as large a vent area as possible, and allow unrestricted venting of combustion gases. Early vent deployment requires that the vent release at the lowest possible pressure ... usually slightly larger than the highest expected differential pressure associated with wind loads—typically 0.96 to 1.44 kPa (0.14 to 0.21 psig)."

If a building's explosion vents conform to industry standards (e.g., NFPA 68), it may be possible to lower the probability of explosion/detonation to account for the vents. However, the issue should be evaluated carefully before giving credit for such measures. It has been shown that the NFPA methods are not always sufficient (Thomas et al., 2006), for example, when a high-speed flame is generated in a space with significant obstacle arrays.

1.5.2   Explosion Suppression

An incipient explosion may be prevented by rapidly introducing a suppression agent into the (generally enclosed) space. Again, credit may be given for such measures in explosion probability calculations if the suppression can be demonstrated as being effective.

1.6   DETONATION OVERVIEW

Detonation of a flammable mixture may occur either directly, through a very strong ignition source, or indirectly, via a deflagration-to-detonation transition (DDT). These are very different and are discussed briefly below. However, the prediction of a detonation probability is beyond the scope of this book due to the many physical and other complexities associated with the issue.

Note that the *detonation* limit of a chemical may be different than the flammable or explosive limits. However, detonation ranges reported in the literature should not be relied upon because more recent testing has shown that the ranges vary significantly depending on the equipment used to perform the test.

1.6.1   Detonation Using a Strong Ignition Source

In this type of event, detonation is initiated by the presence of a very strong ignition source—one that provides many thousands of joules of energy or more, in contrast to simple ignition, which often occurs at less than 1 mJ. In a typical process plant environment, this type of ignition source should only be present as a result of gross failure of significant electrical equipment.

1.6.2   Deflagration-to-Detonation Transition

This phenomenon is an extreme form of the flame acceleration process that leads from a deflagration to a detonation but is qualitatively different, as described in detail in Babrauskas (2003) and more concisely in CCPS (2010). Although the precise mechanism of DDT is described differently by different observers, it basically involves a combustion process that is non-uniform and in the process develops unstable folds and waves at the flame front that build on each other until a shock wave forms. Two versions of this process, and the resulting instability, are described by CCPS (2010).

1.6.3   Buncefield

The Buncefield explosion [Health and Safety Executive (HSE), 2012] is a case study of an event that for some period of time defied explanation since there is some debate whether a detonation actually occurred or not. The proposed explanations are instructive; however, this type of event is enough of an anomaly that an ignition probability tool of the type described in this book may not be capable of identifying it as a possibility.

1.7   OTHER IGNITION TOPICS—HYDROGEN

Hydrogen is a subject of special interest because of its widespread use, extremely low ignition energy (see also the earlier discussion of MIE), and high fundamental burning velocity (which can promote damaging explosion overpressures). Hydrogen is also unique in that anecdotally people have widely different experiences (or opinions) on the ignitability of hydrogen in a given situation.

1.7.1   Ignition Mechanisms

Remarkably, the mode by which hydrogen ignites is still under debate, and several possible mechanisms have been suggested that are potentially applicable to releases into an open environment:

Electrostatic Ignition—Ignition due to sparks, brush discharges, and corona discharges have each been attributed to hydrogen ignition events. Interestingly, a comment is made in Gummer and Hawksworth (2008): *"Studies undertaken many years ago on hydrogen vents ... showed that ignition was rare during fine weather, but was more frequent during thunderstorms, sleet, falling snow, and on cold frosty nights."*

Reverse Joule-Thompson Effect—Hydrogen is an unusual chemical in that, when depressured, its temperature will rise rather than fall. However, this temperature rise is generally modest and so would normally be insufficient to cause the hydrogen to reach its AIT.

Hot-Surface Ignition—As noted earlier, the ignition of a material by a hot surface typically requires temperatures substantially higher than the AIT. From experimental work with hydrogen, it appears to follow this rule.

Diffusion Ignition—Ignition of a 3:1 mixture of hydrogen and nitrogen at 575 K (well below the experimental AIT for hydrogen) has been reported experimentally in a shock tube environment in which the material reached Mach 2.8. This might be analogous to the conditions during a high-pressure hydrogen leak, although under such circumstances it is not clear whether any ignition would be due to this ignition mechanism or by one of the previous mechanisms. Refer to Wolanski and Wojcicki (1972) for more details.

Adiabatic Compression/Turbulence—In this case, the equipment geometry at or near the point of release drives compression that results in a shock wave, as described in Section 1.3.3.6 by Dryer et al. (2007) and recounted by Hooker et al. (2011). The importance of back-mixing in this mechanism was discussed for ethylene/oxygen systems by Britton et al. (1986).

Gummer and Hawksworth (2008) summarized the work through 2007 by saying that the proposed mechanisms do not account for the reported ignitions (or nonignitions) of hydrogen. They cite some specific incidents (see Section 3.7.3) in which ignitions occurred in an obstructed discharge but not in unobstructed discharges. This is similar to what was observed by Dryer et al. (2007) and Hooker et al. (2011) and was attributed to the relative presence or absence of the formation of turbulence near the point of release and may be symptomatic of the adiabatic compression mechanism above.

Swain et al. (2007) conducted experiments which noted some combinations of hydrogen concentration and discharge velocity in which the mixture was "flammable" but ignition did not occur or was not sustained. This was attributed to insufficient local flame speed to burn back to the source, leading the flame to burn outward until there was insufficient hydrogen available.

Bragin and Molkov (2011) reviewed much of the research above and provide a computational fluid dynamics approach to analyzing the ignition mechanism as it relates to the "shock wave" ignition mechanism.

1.7.2 Other Hydrogen Ignition Topics

An older review of actual industrial hydrogen incidents by Zalosh et al. (1978) is discussed in some detail in Chapter 3. The data could be interpreted as suggesting that the vast majority of hydrogen releases ignite and that about two-thirds of those ignitions are explosive. However, aside from not knowing the environment surrounding the releases that might tend to explosiveness (or not), there is nothing suggested in the report that eliminates the potential for data bias—that is, the propensity for ignition events to be reported more often than nonignitions or for explosions to be reported more often than fires. For this reason, one might consider the Zalosh results to be upper bounds on the ignition/explosion probabilities.

A comprehensive view of hydrogen ignition/safety phenomena is provided by Molkov (2007). In addition to many of the references discussed above, Molkov notes the following observations:

Athough the experimental value of hydrogen MIE is a comparatively very low, 0.017 mJ, "at the lower flammability limit the ignition energy requirement of hydrogen is similar to that for methane." Thus hydrogen may not be as easy to ignite as might be assumed based solely on the reported MIE. One might even make an argument that the effective MIE for immediate ignitions (where the hydrogen concentration is highest) should be treated as being lower than the effective MIE for delayed ignitions (after the hydrogen concentration has dropped nearer to LFL).

"Gaseous hydrogen storage cylinders and vehicle storage tanks with pressures up to 1,000 atm can be characterized by a high probability of the spontaneous ignition ... *if special measures are not undertaken*" (italics added).

There appear to be many mechanisms for hydrogen ignition; ignition itself appears to be highly dependent on environmental factors near the point of release that can be expected to vary widely from one release to the next. This aspect of hydrogen ignition likely explains the wide variety of experiences people have had, but as a result introduces a greater level of uncertainty in the analysis. It does not seem practical to accommodate all the possible variables in a hydrogen analysis, if for no other reason than the circumstances of the release may alter these conditions from the normal physical environment. One might make an argument that a release from equipment that is immutable into an open environment (e.g. release through a clean orifice) is less likely to ignite than similar releases resulting from equipment damage or otherwise have flow paths that result in turbulence-induced shock waves.

2 ESTIMATION METHODS

2.1 INTRODUCTION

This chapter presents algorithms that are proposed for estimating ignition probabilities. These algorithms are presented at three levels of sophistication, with the following anticipated uses:

Level 1 (Basic) Analysis—Suitable for PHA risk matrix applications and possibly for LOPAs and FMECAs.

Level 2 (Intermediate) Analysis—Suitable for LOPAs, FMECAs, and screening-level quantitative risk assessments.

Level 3 (Advanced) Analysis—Suitable for QRAs, related cost-benefit analyses, and detailed consequence studies where the frequency is also estimated.

A user may choose to use a higher level algorithm than suggested above. In many cases the benefit derived may not justify the additional effort. However, particularly in situations that are "atypical" (e.g., extremely high or low chemical MIE values), an increased level may be necessary to achieve a sufficiently accurate result. Users are discouraged from using a lower level algorithm than suggested above except when performing an interim or screening-level analysis that will be revised at a later date.

The algorithms that follow are based on the body of literature described in detail in Chapter 3 as well as expert judgment where there are gaps or inconsistencies in the literature. In addition to the models cited in Chapter 3 and those proposed here, there are several proprietary ignition probability approaches that may be appropriate for users with applications that fall outside the standard scope of this book.

2.1.1 Event Tree

In most cases, a given initiating event can have a number of outcomes depending on the circumstances present at the time of the event. The event tree method is a common approach used to quantify the frequency of each outcome. An event tree is typically read from left to right, starting with an "initiating event" and progressing to a multiplicity of outcomes. Figure 2.1 shows an event tree in a form that would be typically used in a risk assessment.

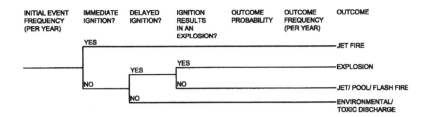

Figure 2.1. Event tree for calculating loss of containment outcome frequencies.

It may be necessary to insert additional branches into the event tree. For example, an additional branch could be added to account for wind direction, since ignition sources, areas of congestion/confinement, and/or receptor buildings may not be present in some directions away from the source. Similar situations are discussed in Sections 1.2.7 and 4.1.3. Additional branches may be needed to account for active protection systems working or failing. Even so, the event tree shown in Figure 2.1 is a useful, basic form that can be modified as needed to accommodate other relevant parameters and that illustrates the basis for the algorithms.

In its simplest form, the event tree does not take into account additional complexities such as meteorology. Frequencies of some of the more significant outcomes can be quantified as follows. Note that throughout this book the term "probability" should be interpreted as "conditional probability"—that is, the probability that an event takes place given the knowledge that a prior event has already occurred. On this basis the following can be quantified:

Frequency of Explosion = Initial release frequency × (1 − Probability of immediate ignition) × (Probability of delayed ignition) × (Probability that delayed ignition results in an explosion)

Frequency of Fire (but no explosion) = Initial release frequency × [Probability of immediate ignition + (Probability of delayed ignition) × (1 − Probability that delayed ignition results in an explosion)]

Frequency of Unignited Discharge = Initial release frequency × (1 − Probability of immediate ignition) × (1 − Probability of delayed ignition)

These are illustrated in Figure 2.2.

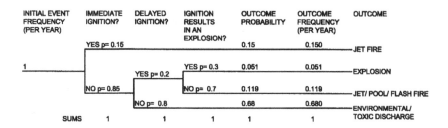

Figure 2.2. Basic quantified event tree example.

Note that if a fire occurs subsequent to an explosion, it must be accounted for separately.

2.1.2   Failure Frequency Data for Use in Event Trees

There are many public and private failure rate databases, and this book does not endorse any single resource. There are very good reasons for some resources to be more appropriate in some applications than others. A thorough review of publicly available failure rate databases is provided in the CCPS book "Guidelines for Evaluating Process Plant Buildings for External Fires, Explosions, and Toxic Releases" (CCPS, 2012b).

2.1.3   Quantification of the Event Tree

The remainder of this chapter is devoted to quantifying this basic event tree for conventional process industry applications using the concepts described in Chapter 1 and the data sources discussed in Chapter 3. There are the typical complications that might be expected—for example, many of the references in Chapter 3 do not make a distinction between the probabilities of immediate ignition and delayed ignition and simply report overall ignition probability. However, these and other challenges can be overcome with reasonable confidence through a careful reading of the sources.

2.2   FACTORS INFLUENCING THE PROBABILITY OF IMMEDIATE IGNITION

Several variables have been identified by one or more sources as being relevant to predicting the probability of immediate ignition (POII). Some of these variables are self-evident—for example, if a material is released at a temperature much higher than its reported autoignition temperature, then it should be expected to ignite. The variables in the discussion that follows are prioritized by the apparent importance of the variable to POII and not necessarily by the number of sources that specifically discuss it, since many of the references only address small subsets of the overall POII question.

There are some types of ignition sources that are beyond the scope of this tool to address. An example is the high ignition probability that would follow a high-energy release event such as a catastrophic equipment failure or impact (e.g., dropped object or pipeline digging accident). Such events are probably better assessed using the judgment of the analysts rather than the algorithms in this book.

2.2.1 Temperature of Release Relative to the Autoignition Temperature

It is assumed in this protocol that autoignition is simply a function of the process temperature and the autoignition temperature of the material being released; that is, there is no external hot surface or flame present at the point of release. As noted in Chapter 1, the autoignition temperature of a material in an open environment is much different (usually lower) than the laboratory test. Yet the risk-based inspection (RBI) protocol published by the American Petroleum Institute (API) assumes that autoignition is not likely unless the release temperature is more than 80° F above the AIT, at which point autoignition is assumed to be a certainty (probability = 100%). This apparent inconsistency may reflect some expectation of cooling upon release.

Because spontaneous fires have been observed at temperatures below the AIT (e.g., when contacted with high-surface-area material such as insulation) and in other cases failed to autoignite when above the AIT, Moosemiller (2010) uses a more nuanced approach. In that paper, it is proposed that spontaneous ignition will never occur below 0.9 times AIT (when AIT is measured in degrees Fahrenheit), will always occur above 1.2 times AIT, and will occur at probabilities between 0 and 1 when the material is released between 0.9 and 1.2 times AIT. Within this range, the proposed probability of autoignition is

$$P_{AIT} = 1 - 5,000e^{-9.5(T/AIT)} \qquad (2\text{-}1)$$

where T and AIT are in degrees Fahrenheit. The latter relationship appears to be more encompassing, although it could reasonably be modified to exclude autoignitions below the AIT if that behavior is not expected.

Note that this equation fails at temperatures below 0 °F. Note too that this relationship and many others in this book are designed to meet expectations and do not have a direct basis in theory. Ignition probability is not a fundamental property and so there is no need or expectation for temperatures, for example, to be expressed in absolute terms such as degrees Kelvin.

2.2.2 Minimum Ignition Energy (MIE) of Material Being Released

Several sources suggest MIE as the most important parameter to evaluate in assessing the "forced ignition" of a flammable release, although almost all discuss POII in terms of specific chemicals rather than in terms related specifically to

MIE. This approach had advantages from the original source point of view, since characterizing the POII in terms of specific chemicals allowed all chemical properties to be "bundled" into a single POII output.

The tools developed in this book must deal with a large number of materials, and so fundamental chemical properties, rather than a list of the chemicals themselves, must be invoked. POIIs that have been suggested for specific chemicals can provide some guidance while acknowledging that chemical properties other than MIE are involved. Some comparisons are made in Table 2.1.

The values in Table 2.1 suggest a relationship where the POII is inversely proportional to the MIE raised to the 0.6 power. This can be compared to the approach proposed by Moosemiller in which the POII (assuming no autoignition) is inversely proportional to the MIE raised to the 2/3 power (in Eq. 2-2, MIE is in mJ and P is in psig):

$$\text{POII (if no autoignition)} = 0.0024 \times (P)^{1/3} / (MIE)^{2/3} \qquad \textbf{(2-2)}$$

**Table 2.1. Proposed probabilities of immediate ignition
as a function of material released**

Chemical	Approx. MIE (mJ)	Proposed POII	Source of POII
Hydrogen	0.015	0.25[a]	Jallais
C1-C2	0.27	0.1	API RBI
Gas from offshore production	0.25	0.1	Spouge
C3-C4	0.26	0.05	API RBI
C5 (liquid)	—	0.02	API RBI
C5 (vapor)	0.22	0.05	API RBI
C6-C8 (liquid)	0.25	0.02	API RBI
C6-C8 (vapor)	0.25	0.05	API RBI
C9-C12 (liquid)	1	0.01	API RBI
C9-C12 (vapor)	1	0.02	API RBI
C13-C16	Assume 10	0.01	API RBI

[a]Represents a combination of several models for specific release conditions. See Chapter 3 for details.

Equation 2-1 assumes that all else is equal with respect to any other variables that may be present for all chemicals. While that is almost certainly not the case (i.e., the average pressure at which hydrogen is handled may be different than the average pressure at which heavy hydrocarbons are handled), this relationship between MIE and POII is a practical and necessary input given the support in the literature for the importance of MIE.

As noted in the Chapter 1 discussion of hydrogen, this particular material is the subject of a great deal of uncertainty and widely varying anecdotes regarding its ignitability. The analyst is encouraged to use company-specific knowledge rather than the algorithms in this book where there is enough history to justify replacing them.

2.2.3   Pyrophoricity of Released Material

If a material is highly pyrophoric, it can be presumed to ignite upon release.

2.2.4   Pressure/Velocity of Discharge

The pressure (and hence discharge velocity) of a release has been cited by several sources as being relevant to the POII. The researchers who commented on pressure/velocity being important attributed this relevance to one or more of the following competing effects:

- Higher pressure results in greater static electricity generation (Swain et al., 2007), although it is noted that there may not be a charge for vapor releases unless they are accompanied by particulates (see next bullet).

- Higher pressure/discharge velocities can generate more fines (e.g., pipe scale), or particulates in the case of liquid releases, that might serve as ignition sources (Pratt, 2000).

- Discharge velocities that exceed a certain rate (possibly the flame speed) cause the flame to "blow off," in which case immediate ignition is not observed (Swain et al., 2007; Britton, 1990a). However, different researchers have had different results, with some reporting ignitions at very high pressures and others reporting blow-off at lower pressures.

- There may be cases where pyrophoric compounds form at the exit point of a release and could serve as an ignition source. However, if these compounds exist, they are only effective at very low pressures (observed at the end of a release) when air is able to migrate back to them. For the purposes of this book, it is assumed that at the pressures required to get ignition by this mechanism the pressures are also low enough that the release is not of interest to the users of this book.

Most of these studies were conducted with hydrogen, and their applicability to other materials is not clear. Moosemiller (2010) included the 1/3 power pressure relationship assuming that it applied for all materials, as indicated in Eq. 2-2, although the basis for that relationship was based more on anecdotal information than "hard" data. See also the droplet size discussion in the next section.

The U.K. Energy Institute (UKEI) also discussed self-ignition and adopted a correlation used in earlier works that stated that the probability of self-ignition is proportional to the mass release rate—that is, proportional to the square root of pressure in nonflashing cases. One can correctly argue that the release rate is related to hole size as well as pressure; this is taken into account in the delayed ignition algorithms in this book but not in immediate ignition as the static effects of pressure are assumed to dominate.

2.2.5   Droplet Size

Droplet size of liquid sprays is considered relevant by Lee et al. (1996), Britton (1999), and Babrauskas (2003) in the sense that it affects the MIE according to a power relationship MIE α D^n, where "D" is the droplet diameter. Different sources put the value of "n" between 3 and 4.5. In the range of 40–150 μm, the droplet size appears to follow the following relationship to MIE based on Babrauskas data:

$$\text{MIE } \alpha \text{ (droplet diameter)}^{3.4} \tag{2-3}$$

Britton (1999) cites the "Simple Chemically Reacting System" model that relates MIE to the cube of the particle diameter.

The droplet diameter is not immediately available to the analyst but is inversely proportional to the release pressure in an ideal case. Thus, the MIE may be assumed to have the relationship

$$\text{MIE } \alpha \text{ (release pressure)}^{-0.3} \tag{2-4}$$

The data of Lee suggest that there is some lower diameter limit to the **relationship described in Eq. 2-3, which makes sense given there is some point at** which a mist would be so fine that it would effectively act like a vapor. A best estimate of this based on comparing liquid and vapor MIEs is that the relationships above are valid down to a droplet size of about 20 μm, below which the mist exhibits vaporlike behavior.

For a large hole of the kind of interest to this effort, very large pressures (several thousand psi) are required to achieve uniform droplets under 20 μm in diameter for liquid phase releases. For the purposes of the protocols being developed in this book, it is therefore assumed that the MIE of a spray is equal to the MIE of the vapor for the material being released multiplied by a factor related to a baseline of 20 μm/10,000 psig, that is,

$$MIE_{liquid} = MIE_{vapor} \times (10,000/P)^{0.3} \qquad \text{(2-5)}$$

It is also possible to create micrometer-sized droplets that are ignitable by sudden cooling of a supersaturated vapor, although this phenomenon is outside the scope of this book to address.

2.2.6   Presence of Particulates

Multiple authors have proposed particulates as a source of immediate ignition in flammable releases. These particles are presumed to be generated during the flammable release (e.g., scale from piping). One company (private communication) has also noted that particulates can be present in the process and released with flammables—particularly if the containment equipment fails at the time of the flammable release. It is not clear whether the level of particulates can ever be predicted for a given event, and to a large degree the discharge of particulates may be related to variables such as pressure that are already considered. For these reasons, there is no specific POII modifier suggested for the particulate issue.

2.2.7   Configuration/Orientation of Equipment Near/At the Point of Release

Some investigators of hydrogen in particular (Dryer et al., 2007; Duarte et al., 1998; Britton, 1990a; Gummer and Hawksworth, 2008) have noted that the ignition probability can be influenced by the configuration of equipment at or near the point of release. Two competing mechanisms are described:

- Narrow points of release resulting in turbulence and/or formation of a shock wave that can result in ignition with any air that is present.
- Very long/narrow discharges can prevent ignition due to escape of heat from any incipient flame front.

This effect may be of interest in the smallest of releases (e.g., flange leaks). However, these smaller releases are of less interest to the scope of this book; therefore, while the phenomenon may exist, it is not accounted for here.

2.2.8   Temperature of Release (As It Relates to Its Effect on MIE)

Babrauskas (2003) reports the effect of temperature on MIE, as described in Chapter 3. The effect on vapor releases is less pronounced than for liquid releases:

$$MIE_{vapor} = MIE_{vapor} \text{ at } T_{ref} \times \exp[0.0044(T_{ref} - T)] \qquad \text{(2-6)}$$

$$MIE_{liquid} = MIE_{liquid} \text{ at } T_{ref} \times \exp[0.016(T_{ref} - T)] \qquad \textbf{(2-7)}$$

where temperatures are reported in degrees Fahrenheit.

If this effect is accounted for in releases that undergo large changes in pressure or temperature, the temperature at the ignition source should ideally be used rather than the process temperature.

2.2.9   Phase of Release (API RBI)

The API RBI protocol, based on expert judgment, treats liquid and vapor POIIs of the same materials differently, generally ascribing a POII for a vapor release about three times higher than the POII for a liquid release. Since there is no specific data cited to support this opinion, it is assumed that the release phase is more accurately treated by using phase-specific values for the MIE.

2.2.10   Flash Point and Release Rate (TNO)

TNO produced a POII table based on these inputs. However, as they are not mentioned by other investigators for immediate ignition, and can mostly be related to other parameters such as pressure that are already incorporated, they are not utilized separately here for POII.

2.3   FACTORS INFLUENCING THE PROBABILITY OF DELAYED IGNITION

As was done for the probability of immediate ignition, the major variables affecting the probability of delayed ignition (PODI) are discussed next. The initial discussions will be on the basis that the release is taking place outdoors; indoor releases will be addressed near the end of this section.

2.3.1   Strength and Numbers of Ignition Sources

Many investigators have commented on the relevance of ignition source "strength" and "density." While the concept of "density" of ignition sources is straightforward, the use of "strength" in ignition probability literature is more intuitive than reducible to objective terms such as electrical potential or energy. The ignition strength has frequently been expressed either in terms of point sources or in terms of area sources. The point-source approach is difficult to apply if there are a number of sources, since one cannot simply add up the contributions of each individually without accounting for the likely ignition by one of the other sources.

Mathematically, this can be described simply as

$$P_n = P_{n-1} + p_i(1 - P_{n-1})$$ (2-8)

where P_n is the PODI for the combined "n" ignition sources being evaluated and p_i is the PODI of an individual ignition source in the absence of any other ignition sources. However, this simple relationship belies the complexity of performing the calculation for a large number of ignition sources.

Because of the complexity in dealing with hundreds of potential ignition sources that may be present in a process plant, the area-source or line-source approaches are also frequently used. Chapter 3 provides estimated ignition strengths for a variety of point and area ignition sources; a summary of the various references is provided in Table 2.2.

Note that the values in Table 2.2 assume that the flammable cloud is in contact with the ignition source. Thus, for example, a ground-level release may almost always ignite in the presence of a ground flare ("point sources/flare" strength = 1.0 in Table 2.2); however the same release will probably never be ignited by an elevated flare, in which case that ignition source should not be considered.

In a similar manner, a well-defined flammable cloud may enter a congested process area and yet not encounter a significant number of ignition sources. This can occur, for example, if the release is elevated and remains so. An example of this is illustrated in Section 4.3.5. In such a case, the analyst may choose to "downgrade" the availability of ignition sources, e.g., from "medium-density process" to "low-density process." However a decision to do so should be documented to assure consistency in approach from one application to the next.

Also note that these proposed values are averages across a broad spectrum of release chemicals and situations, and individual location experience may be different. One site, for example, has had ignition occur both times a flammable cloud contacted a high-power line. This could be indicative of something unique about that particular power line, or the material being released, or simply bad luck.

The ignition source strengths in Table 2.2 are commonly expressed in terms of the probability of ignition if the flammable cloud is exposed to the ignition source for 1 minute; however, this is not always the case. The actual ignition probability during a 1-minute period will also be dependent on other factors, chiefly the ease of igniting the material being released. Given the common usage of ignition "strength" in the literature, however, the 1-minute basis for ignition "strength" values is a reasonable if broad frame of reference for discussion purposes.

Table 2.2. Ignition source "strength"

Ignition Source Type	Ignition Source	Strength "S"	Source
Point Sources	Fired heater	0.9	Moosemiller
	Boiler (outside)	0.45	TNO
	Boiler (inside)	0.23	TNO
	Flare	1.0	TNO, HSE
	Motor vehicle	0.3	TNO
	Ship	0.4	TNO
	Diesel train	0.4	TNO
	Electric train	0.8	TNO
	Hot surfaces	See discussion below table	
Line Sources	High-power electrical line	0.001 x length of line covered by cloud (ft)	TNO, Moosemiller
	Roadway (if area of cloud is known)	$1 - 0.7^V$ (V=average number of vehicles covered by the cloud)	Moosemiller
Area Sources	Process unit (if area of cloud is known)	0.9 x fraction of unit covered by the flammable cloud	Modified TNO
	Residential population (if area of cloud is known)	$1 - 0.99^N$ (N=number of people covered by the cloud)	Modified TNO, Moosemiller
	High-density process area (outdoor)	0.25[a]	UKEI/HSE
	Medium-density process area (outdoor)	0.15[a]	UKEI/HSE
	Low-density process area	0.1	UKEI/HSE
	Confined space with ~ no equipment	0.02	Moosemiller
	Process area outdoor storage	0.1	HSE, Moosemiller
	Remote outdoor storage area	0.025	Assumed, relative to previous row
	Office space	0.05	UKEI

[a]The original source provided values for indoor operations. These have been scaled down by roughly a factor of 2 for outdoor operations.

Hot Surfaces—A few investigators (API, 2003; Hamer et al., 1999; Duarte et al., 1998) have looked at the issue of hot surfaces as sources of autoignition after release. Hot surfaces may be present as a result of operating high-speed equipment (e.g., motors), hot process piping (e.g., high-pressure steam), hot reactors (e.g., oil refining fluid catalytic cracking catalyst regenerator), and more. Note that there must be sufficient available hot-surface area to sustain the heat given the prevailing air flow around the surface; thus, equipment such as motors might be the smallest practical hot surface described by this tool. Duration of exposure and velocity over the surface are also important. API data suggests that ignition typically occurs near the AIT after about 1 minute of exposure to the hot surface; shorter exposure times and increased wind speeds require much higher temperatures. The latter phenomenon is presumably because the material passing over the hot surface does not have enough time to reach its AIT.

The duration effect is addressed as for other ignition sources, which is discussed shortly. The strength of a hot-surface ignition source analogous to those in Table 2.2 is proposed to be expressed as follows:

$$S = 0.5 + 0.0025[T - AIT - 100(CS)] \qquad \text{(2-9)}$$

where T (the temperature of the hot surface) and AIT are expressed in degrees Fahrenheit and the cloud speed (CS) is expressed in meters per second. Note that the cloud speed would be the same as the reported wind speed for clouds that have passed any jet portion of the release but may be substantially higher if the hot surface is exposed to the jet portion of the release.

A minimum of zero is set on the S value obtained using this equation.

Control of Ignition Sources—Some authors have incorporated the effectiveness of ignition controls in one guise or another. Spencer et al. (1997, 1998) provide different ignition source strengths for controls that range from "none" to "ideal" and from "poor" to "good." The difference between "good" and "poor" is typically a POI factor of 2, although for some ignition sources the differences are more extreme. Daycock and Rew (2004) define the following categories of controls:

"Typical"—the level of control that is considered to be the most realistic for a typical hazardous installation

"Good" —the highest level of control that is practical

"Poor" —the minimum level of control that may occur in practice

These terms are subjective to some degree, so the user should be able to define these terms in the context of their operations sufficiently well to ensure consistency in application from site to site. Although these terms are subjective, this book permits their use in the spirit of providing motivation for improvement. However, the difficulties of objectively evaluating the level of control at a point in time *and at the future time that a release occurs* is fraught with uncertainty. Therefore a limit of a factor of 2 is proposed between "good" and "poor."

2.3.2   Duration of Exposure

Typically coupled with the discussion of ignition source strength is the issue of exposure duration. Most investigators have adopted an ignition source strength and duration relationship of the following form:

$$P_{delign} = 1 - k \times e^{-at} \tag{2-10}$$

where k is a strength constant, a is a time constant, and t is the exposure time. After reviewing the literature on the subject, Moosemiller (2010) concluded that both k and a were related to the ignition source strength as follows:

$$k = 1 - S^2$$

$$a = S$$

Therefore,

$$P_{delign} = 1 - [(1 - S^2) \times e^{-St}] \tag{2-11}$$

where t is expressed in minutes. This is a much simpler form of the relationships developed by Spencer et al. (1997, 1998) but more practical for the algorithm types and applications developed for this book.

2.3.3   Release Rate/Amount

Numerous investigators have commented on the relationship between release rate and ignition probability. This is not surprising—if nothing else, a larger release should produce a larger cloud that can contact more ignition sources. In their seminal work, Cox et al. (1990) developed data-based mathematical relationships relating PODI to release rate for offshore liquid and gas phase events that are still widely used.

Ronza et al. (2007) considered releases from transport vehicles and expressed the probability of ignition (POI) as a function of the total amount spilled. Again, larger spills had higher probabilities of ignition.

The relationships developed by others generally take the form

$$POI = a \times (Release)^b \tag{2-12}$$

**Table 2.3. Power law constants for relationship of
release magnitude to probability of ignition**

Source	Basis	Power Law Constant "b"
Cox et al. (gas)	Rate	0.642
Cox et al. (liquid)	Rate	0.393
Ronza et al.	Volume	0.3
Crossthwaite et al.	Rate/volume	0.5

Some constants that have been proposed are listed in Table 2.3.

Curiously, other more recent offshore data (Thyer, 2005) suggests a negative correlation of release rate to ignition probability. This may reflect more rigorous management of ignition sources in the years since the Cox et al. data or some qualitative difference between onshore and offshore operations (which is why this book is not directed at offshore applications). Regardless, all else being equal, the relationship between release rate and ignition probability might be expected to be less pronounced for offshore facilities than onshore facilities because there is a finite footprint over which ignition sources exist offshore.

In the face of some inconsistencies but recognizing that a larger release should result in a larger flammable cloud (at least for releases not contained by a dike, etc.), the following relationships are proposed:

$$\text{PODI (liquids)} \; \alpha \; (\text{Release Magnitude})^{0.3} \qquad \textbf{(2-13)}$$

$$\text{PODI (vapors)} \; \alpha \; (\text{Release Magnitude})^{0.5} \qquad \textbf{(2-14)}$$

Since the release rate varies with the square of the hole diameter, this can be related to the hole size as follows:

$$\text{PODI (liquids)} \; \alpha \; (\text{Hole Diameter})^{0.6} \qquad \textbf{(2-15)}$$

$$\text{PODI (vapors)} \; \alpha \; (\text{Hole Diameter}) \qquad \textbf{(2-16)}$$

2.3.4   Material Being Released

For many of the same reasons discussed for POII, the material being released has an obvious effect on the PODI. First, the volatility of the material released will impact its dispersion distance. Second, once the material reaches an ignition source, the ease with which it ignites will determine the probability of ignition.

To some extent, the volatility issue will be addressed by the user's ignition source inputs—if the cloud cannot reach an ignition source, that ignition source should not be considered. (Note it is beyond the scope of this book to provide guidance on dispersion modeling.) Since both the dispersability of a cloud (as expressed by the material's vapor pressure, etc.) and its propensity to ignite (as expressed by MIE) are strongly correlated, it is desirable in this case to couple these factors into a single parameter. MIE is chosen for this purpose.

API RBI (API, 2000) has developed PODI tables that can be used as the basis for defining a MIE/PODI relationship. This relationship can be roughly described as follows:

$$PODI \sim 0.02 - 0.23 \log(MIE)$$

Moosemiller (2010) has also developed such a correlation, which can be rendered as follows:

$$PODI \sim 0.18 - 0.26 \log(MIE)$$

This book supports an alternative that tends toward predicting higher ignition probability for typical situations:

$$PODI = 0.15 - 0.25 \log(MIE) \qquad \text{(2-17)}$$

The PODI determined using this equation will have a maximum value of 1 and an assumed minimum value of 0.001.

2.3.5   Release Phase/Flash Point/Boiling Point

The phase in which a material is released (or ultimately becomes after release) has been mentioned by many investigators as being relevant to POI. However, the reasons for the apparent differences in liquid and vapor probabilities of ignition are reflected in the factors mentioned earlier, such as MIE, pooling of liquids (limiting surface area for vapor generation), and so forth. Therefore, no new parameters are suggested to account for the release phase, except to the extent that the phase affects the MIE, as discussed in Section 2.2.8.

However, the boiling point of a material is related to other parameters noted in Chapter 1 as being relevant—flash/fire point, vapor pressure, and indirectly the flammable limits. All else being equal, a pool of high-boiling-point material is expected to have a lower probability of ignition than a pool of low-boiling-point material.

It is also likely that the boiling point is related to the MIE or AIT, at least within specific chemical groups, although this is not always true. This raises the potential for any boiling point modifier that might be introduced to unintentionally add onto other modifiers that are influenced by the boiling point as well. Therefore, its importance must not be overstated.

There are other competing factors. For example, the probability of ignition might, in principle, be connected to the area or volume of a flammable vapor cloud. But this relationship is not likely to be linear since, for example, more remote ignition sources will have less opportunity to ignite a cloud than ignition sources that are closer, since ignition by a closer source precludes ignition by the more distant source.

In sum, it is convenient to assume that the probability of ignition is related to the boiling point or flash point of the material relative to its release temperature. Use of the boiling point as the reference material property can be more directly linked to other chemical and dispersion properties, but flash point is a more direct measure of the ability of an ignitable cloud to form. Therefore inputs are proposed later in this chapter using either input.

2.3.6   Distance from Point of Release to Ignition Source

The distance between the point of release and the ignition source is an obvious factor but is considered to be accounted for by other inputs (e.g., release magnitude) and user judgment (the expectation that someone would not enter an ignition source that could not be reached by the release). Therefore, it is not considered further in this book.

2.3.7   Meteorology

The prevailing weather conditions affect the PODI to the extent that they impact the dispersion of the cloud. The humidity is relevant to some degree but is not often described in the literature as being important except for indoor operations involving chemicals with high electrostatic potential. The wind speed has also been noted previously as being relevant to ignition by a hot surface.

Each of these features is implicit in one of the other parameters discussed or is beyond the scope of this book. Therefore, meteorology is not considered except as previously mentioned and in the discussion of indoor operations that follows.

2.3.8   Events Originating Indoors

A few investigators have considered the potential for indoor flammable ignitions compared to outdoor ignitions and have commented on the qualitative differences between the two environments. There is no known data to support separate quantification of indoor events, but given that there is a belief that they are different and are governed by different variables, it is useful to consider expert opinion in the absence of actual data.

The key variables of interest include:

- Material released (volatility, ignitability, electrostatic properties)
- Release rate vs. ventilation rate
- Presence of gas detectors and activation/response strategy
- Humidity
- Fraction of time the space is active
- Ignition prevention measures that are in place

2.3.8.1   High-Level Analysis

Based on consideration of release confinement and persistence, Moosemiller (2010) proposed using a simple multiplier of 2 to account for indoor operations vs. equivalent outdoor operations. Of course, the true multiplier is highly dependent on the controls in place, such as electrical classification, ventilation, etc.

2.3.8.2   Material Released

In most respects, it will be assumed that the material properties are addressed in the same manner as those for outdoor releases. For some indoor operations, there is an additional potential for ignition due to human static discharge. This probability is obviously related to the MIE (discussed earlier) and will be treated as such. The potential for human ignition sources is in addition to other indoor sources.

2.3.8.3   Human Ignition

People can act as ignition sources even if other indoor sources have been effectively eliminated. Ignition by electrostatic discharge is reported to be more difficult to achieve than what the laboratory MIE might suggest. Johnson (1980) and others have noted that an ignition source much greater than the MIE is required for human electrostatic ignition. Johnson found that ignition could occur with an electrostatic discharge as low as 2.4 × MIE; others have suggested a discharge in the range of 60–100 × MIE.

Johnson listed the following additional relevant factors:

- Presence of the vapor/air mixture within a narrow concentration range
- Sufficient electrostatic charge generated on an operator
- Storage of the charge on the operator for a sufficient period of time
- Discharge to a large or grounded object

It is assumed that the probability of ignition for indoor release is approximately twice as high as those for outdoor releases due to the enclosed spaces, absent of any corrections for ventilation. It will also be assumed that human electrostatic ignition will not occur at an electrostatic discharge less than *twice* the MIE

Narrow Concentration Range—An example of just how narrow the concentration range needs to be for human-based ignition is illustrated in Chapter 3. Such conditions will be assumed to exist at the location of the ignition source (person) in 20% of all events, based on some expected movement during the event.

Charge Generation—During a release event, an operator can generate an electrostatic discharge through any number of activities. It is assumed that these conditions are always present.

Charge Storage—According to Johnson, a relative humidity of 60% at 70°F is sufficient to prevent human static ignition because the charge will leak to the ground due to increased surface conductivity. Therefore, assuming that indoor operations generally take place at ~ 70°F, the following is assumed: $PODI_{human}$ α [100 – 1.5 × relative humidity (%)] for humidity values under 60%.

Spark Discharge—It is assumed that a spark discharge will always be possible unless antisparking floors are used. If antisparking floors are present, then a spark discharge will be assumed to be impossible.

2.3.8.4   Release Rate vs. Ventilation Rate

The difference between the release rate and the ventilation rate affects the relative probability that a cloud in the flammable range will be present or persistent. Actual concentration profiles can be calculated relatively easily assuming that ventilation air mixes instantly with the existing air and released chemical. Much better tools such as those based on computational fluid dynamics (CFD) are available.

CFD or similar methods can be used to determine the concentration profile within a building. However, just as with outdoor releases, there is no direct correlation between this concentration profile and the ignition probability. There is only one reported attempt in the literature to quantify this relationship, and that is the "speculative" model proposed by Moosemiller. This model is discussed in Appendix B.

2.4   FACTORS INFLUENCING THE PROBABILITY OF EXPLOSION, GIVEN DELAYED IGNITION

This issue has been discussed only infrequently in the literature, perhaps due to the fact that the propensity to explode is controlled primarily by plant layout considerations that are usually out of the control of the analyst to influence.

The potential for explosion is driven by the following factors:

- The level of congestion and confinement in the area in which the cloud is ignited
- The propensity of the fuel to ignite explosively (fundamental burning velocity)
- The presence or absence of explosion relief or suppression systems

The first two of these factors can be correlated to a blast wave velocity using various approaches that are described in other CCPS books (e.g., CCPS 2010). This blast wave velocity then defines whether an "explosion" has taken place. The effects of congestion and confinement on the potential for explosion (and the degree of severity) are highly dependent on spatial considerations that are beyond the scope of this book to address. Therefore, no rigorous model for explosion probability is presented in this book. A model is provided in Appendix B that may be considered for estimating the probability of an explosion taking place *given the necessary preconditions of congestion and confinement* as determined through consequence modeling.

2.5   POTENTIAL INTERDEPENDENCE OF VARIABLES

It can be argued that there is the potential for variables and the algorithms in this book to interfere with related tools that are outside the scope of this book. This occurs most obviously with inputs that have a relation with the consequences of a release. One of these issues is the levels of congestion and confinement that are present in the flammable cloud when it is ignited, as discussed in Section 2.4.

Other potential complicating factors include the following:

Release Rate—Several investigators have noted the relationship between release rate and ignition/explosion probabilities, and this input is incorporated in the models in this book. However, the same variable is utilized in consequence modeling which could, in principle, be used to adjust the probability of ignition through estimating the probability that a cloud would contact a specific ignition source. Thus there is the potential to underestimate the probability of ignition by multiplying the probability as calculated in the tool and the probability resulting from a consequence model or event tree (wind direction) outside the tool. It is also important to note that the data sources from which the algorithms are derived have the same convoluting factors implicit within them—or, more accurately, some probably do and some probably do not—and that the probability of ignition as estimated using the tools in this book incorporates many more variables than simply special considerations. It is therefore suggested that analysts who want to incorporate aspects of both consequence and probability models into their probability analysis should set the event magnitude inputs used in this tool to the values that make the resulting magnitude multipliers equal to 1.

Release Temperature Relative to Normal Boiling/Flash Point—This input, introduced in Section 2.3.5, is intended to provide a factor that describes the size of a cloud and hence its potential to reach multiple ignition sources. This has an obvious potential to conflict with, or be redundant with, the results of a dispersion model. Therefore, if a user has dispersion model results that are to be incorporated in ignition probability logic, the temperature input should be set to be equal to the normal boiling point of the material so that the temperature modifier (as described in Section 2.8.2.5) is set to 1.

2.6 SUMMARY OF VARIABLES USED IN EACH ANALYSIS LEVEL

Table 2.4 summarizes each variable used in the algorithms, and the level(s) in which it is used.

Table 2.4. Summary of variables used in each analysis level

Model Result	Variable	Basis/Comments	Levels
Prob. of Immediate Ignition	Process Temperature T	See next row. In Levels 2 and 3, T is also used to adjust the effective MIE value. In Level 3, the temperature at the release outlet can be substituted for the process temperature.	1,2,3
	Autoignition Temperature AIT	Ratio of T to AIT sets probability of autoignition.	1,2,3
	Process Pressure	Increasing pressure has been identified as contributor to static charge formation.	2,3
	Minimum Ignition Energy MIE	MIE is a measure of the ease with which a material will ignite.	2,3
Prob. of Delayed Ignition	Release Indoors or Outdoors	Presumption that confinement of release allows greater persistence to contact ignition sources. See also "Ventilation" and "Human" discussions below for Level 3.	1,2,3
	MIE of Released Material	MIE is a measure of the ease with which a material will ignite.	1,2,3
	Ignition Source Strength S	The greater the energy in the ignition source, the greater the chance of ignition.	2,3
	Release Duration	The longer the release is exposed to the ignition source, the greater the chance of ignition.	2,3

Table 2.4 (continued)

Model Result	Variable	Basis/Comments	Levels
	Release Magnitude	Various investigators have noted that larger releases have a higher probability of ignition.	2,3
	T, and Normal Boiling Point	For liquid releases, this describes the ease with which a release will become a vapor.	2,3
	Control of Ignition Sources	Investigators have proposed that sites with better-than-average control of ignition sources have lower ignition rates than those with average or poor controls.	3
	Ventilation and Detection	Indoor facilities with rapid leak detection and ventilation capabilities should have lower POIs than those that do not.	3
	Human Influence	Investigators have noted that indoor ignitions are less likely in high-humidity environments.	3
	Mitigation	User allowed manual entry of measures that would prevent ignition, such as inerting, deluge, etc.	3

2.7   BASIC (LEVEL 1) PROBABILITY OF IGNITION ALGORITHMS

2.7.1   Level 1 Algorithm for Probability of Immediate Ignition

2.7.1.1   Contribution of Static Ignition

Typical immediate ignition probabilities reported in the literature are on the order of 0.05 (Crossthwaite et al., 1988—for LPG) or less (UKEI, 2006—for oil and gas production) and are assumed to be based predominantly on the contribution of static ignition rather than autoignition since most processes operate below the AIT. Improving this estimate involves input effort that is beyond the expectations of a Level 1 analysis, and so for Level 1 the static ignition contribution is assumed to be 0.05.

2.7.1.2   Contribution of Autoignition

Relationships between the process temperature relative to AIT and the POII have been developed as described earlier. The relationships chosen here are

$$\text{If } T/AIT < 0.9, \text{ then } P_{ai} = 0 \tag{2-18}$$

$$\text{If } T/AIT > 1.2, \text{ then } P_{ai} = 1 \tag{2-19}$$

$$\text{If } 0.9 < T/\text{AIT} < 1.2, \text{ then } P_{ai} = 1 - 5,000e^{-9.5(T/\text{AIT})} \qquad \textbf{(2-20)}$$

with the temperatures in degrees Fahrenheit. For pyrophoric materials, the probability of immediate ignition is assumed to be 1.0. Note that the relationships above do not yield valid results for chemicals with extremely low (e.g., subzero) AITs. These chemicals are rare enough that they should probably be handled on an experiential basis in any case.

2.7.1.3   Combined Level 1 Algorithm for POII

The combination of the logic developed in the static and autoignition discussions above is the following, after avoiding interaction between the two:

$$\text{POII}_{\text{Level 1}} = 0.05 + (1 - 0.05) \times P_{ai} \qquad \textbf{(2-21)}$$

with P_{ai} calculated as above.

This analysis only requires the input of the material being released (that is, its AIT), and the temperature in the process. Note that it is possible to suggest cases in which the POII calculated above would be 1 but which under certain circumstances might not ignite. A POII of 1 is potentially nonconservative with respect to the potential probability (and impact) of a delayed ignition. For this reason, the value of POII is limited to 0.99 in order to preserve the delayed ignition portion of the event tree.

2.7.2   Level 1 Algorithm for Probability of Delayed Ignition

A goal of a Level 1 algorithm is that it uses a minimum of readily available information. The Level 1 calculation for POII assumed that a pick list of chemicals was available, based on the analysis of the associated chemical AIT and the process temperature. Or, if the chemical was not on a pick list, it would be apparent whether the process temperature was close enough to the AIT that the analysis had to incorporate AIT even if it meant looking up the AIT value.

The drivers for PODI, while significant, are more subtle than the factors that influence POII. Two additional inputs are suggested for the calculation of PODI. The first of these is whether the release is indoors or outdoors. The second of these is the chemical being released.

Equation 2-17 provides the basis for predicting PODI as a function of a chemical's MIE. In comparing expectations from various investigators, a value that errs on the side of higher PODI for an outdoor event is presumed to be 0.25. On that basis, a relatively low default MIE of 0.2 mJ will be used (resulting in a PODI of 0.25) if the user does not enter a chemical or if the chemical/mixture MIE is not known. If the event is indoors, the calculated Level 1 PODI is further multiplied by a factor of 1.5.

2.8   LEVEL 2 PROBABILITY OF IGNITION ALGORITHMS

2.8.1   Level 2 Algorithm for Probability of Immediate Ignition

2.8.1.1   Contribution of Static Ignition

The Level 1 analysis assumed a constant value of 0.05 for the static ignition contribution. The Level 2 analysis will consider the propensity of static ignition to be a function of the material's reported MIE and the process pressure, P. Based on relationships developed by earlier authors, the static contribution is assumed to be as follows:

$$POII_{static} = 0.003 \times P^{1/3} \times MIE^{-0.6} \tag{2-22}$$

where P is in units of psig and MIE in units of millijoules. Comparison of the resulting POI with experience suggests that there may be an upper limit to the pressure effect. It is currently assumed based on anecdote that 5,000 psig is a reasonable upper limit to be allowed for the pressure input.

Improvements to this basic equation can be made to account for liquid phase releases by considering the effects of pressure and temperature on the MIE. These refinements might normally be considered Level 3 sophistication; however, since the necessary inputs have already been provided at this stage there is no reason not to incorporate them in Level 2 (except perhaps if the calculation is being performed manually).

The first step in the refinement is to convert the MIE of a liquid into what is effectively an equivalent vapor MIE value. This is done using the relationship described previously:

$$MIE_v = MIE_{reported} \times (10,000/P_{liquid})^{0.25} \tag{2-23}$$

where MIE_v is the equivalent vapor MIE, $MIE_{reported}$ is the value of MIE reported in the literature, and P_{liquid} is the process pressure of the liquid (P elsewhere). Next, a temperature compensation term is utilized:

$$MIE_{adj} = MIE_v \times exp[0.0044(60 - T)] \tag{2-24}$$

where MIE_{adj} is the temperature-adjusted MIE and T is the process temperature (as before). The improved algorithm for the static contribution of POII is the following:

$$POII_{static} = 0.003 \times P^{1/3} \times MIE_{adj}^{-0.6} \tag{2-25}$$

A maximum of 0.9 is placed on this result due to limits in the underlying data and to ensure that delayed ignition is considered possible.

2.8.1.2 Contribution of Autoignition

Autoignition is a function of temperature and AIT, and while a refinement to the Level 1 approach can be performed, it requires a level of effort associated more with a Level 3 analysis.

Therefore, the Level 1 approach is repeated for Level 2:

$$\text{If } T/AIT < 0.9, P_{ai} = 0$$

$$\text{If } T/AIT > 1.2, P_{ai} = 1$$

If $0.9 < T/AIT < 1.2$, then

$$P_{ai} = 1 - 5{,}000e^{-9.5(T/AIT)} \qquad \textbf{(2-26)}$$

For pyrophoric materials, the probability of immediate ignition is assumed to be 1.0.

2.8.1.3 Combined Level 2 Algorithm for POII

The combination of the previous two subsections is shown by Eq. 2-27, after avoiding interaction between the two:

$$POII_{\text{Level 2}} = P_{ai} + (1 - P_{ai}) \times POII_{\text{static}} \qquad \textbf{(2-27)}$$

with P_{ai} calculated as above.

This analysis requires the input of the material being released (that is, its AIT and MIE), its phase in the process, and the temperature and pressure in the process. For the same reasons as described for a Level 1 analysis, the POII at Level 2 is limited to a maximum value of 0.99.

2.8.2 Level 2 Algorithm for Probability of Delayed Ignition

2.8.2.1 General Approach

There are more variables than usual involved in this level of analysis—ignition source strength, event duration, hole size, release phase, and MIE. In addition, these variables do not operate independently of each other, as was the case for the "autoignition" and "static" contributors to the POII. As a result, calculating "absolute" values for each input introduces complexities when they are combined. Therefore a series of modifiers will be applied to a baseline PODI that is calculated as a function of ignition source strength and event duration.

2.8.2.2   Strength of and Duration of Exposure to Ignition Sources

Sections 2.3.1 and 2.3.2 described the most common mathematical relationship developed for predicting PODI as a function of the ignition source type and the duration of the exposure to the ignition source (which for the purposes of these algorithms is assumed to be approximately the same as the release duration).

This relationship is

$$PODI_{S/D} = 1 - [(1 - S^2) \times e^{-St}] \tag{2-28}$$

where t is expressed in minutes and examples of S for different ignition source types are provided in the tables and text of Section 2.3.1. The values calculated will then be modified by the values developed in the following two subsections.

Testing has shown that there should be a limit to the time effect, since:

- There is reason to believe that if a release has not ignited after some extended period of time, then it probably never will (whereas an infinite model will drive the ignition probability to certainty).

- For the purposes of human safety risk analysis, there is some point in time after a flammable release in which additional event time is moot, since the personnel potentially affected would have been evacuated from harm's way.

For these reasons, a duration limit of 10 minutes is imposed in the software version of this book. If desired readers can, of course, perform the calculation manually using a different duration assumption if such an assumption better suits the specifics at that site.

There are some notable exceptions to this rule in the historical record (Buncefield, Ufa, Port Hudson, etc.). However, these events are considered exceptional and so cannot be adequately incorporated into the ignition probability methods described in this book.

2.8.2.3   Magnitude of Release

Section 2.3.3 described proposed relationships between release magnitude and the probability of ignition. These are either expressed for the overall amount released (as might be appropriate for an instantaneous event) or based on the hole size and release rate. To express these in terms of modifiers for the purposes of combining with the number developed in the previous section, an assumption of the baseline (that is, the "typical" release magnitude) must be made.

Based on the range of events that have been reported by those describing this variable's effect, the following baseline values are assumed to represent a modifier of 1 (average):

- Average total magnitude of liquid release—5,000 lb
- Average magnitude of vapor release—1,000 lb
- Average hole size of release—1 inch equivalent diameter
- Average pressure of release—100 psig

The PODI modifiers for release magnitude are then:

$$M_{MAG\_Amount\ Released\ (liquid)} = (Amount\ Released/5,000)^{0.3} \qquad \textbf{(2-29)}$$

$$M_{MAG\_Amount\ Released\ (vapor)} = (Amount\ Released/1,000)^{0.5} \qquad \textbf{(2-30)}$$

$$M_{MAG\_Hole\ Diameter\ (liquid)} = (Hole\ Diameter)^{0.6} \qquad \textbf{(2-31)}$$

$$M_{MAG\_Hole\ Diameter\ (vapor)} = (Hole\ Diameter) \qquad \textbf{(2-32)}$$

where the amount released is in pounds and the hole diameter is in inches.

Two limitations are set on these multipliers. It is noted that a magnitude-based adjustment may be a result of an extended duration event—an event whose duration is already accounted for in large part in Eq. 2-11. To minimize multiplicative effects of extended duration events, such that the effects of Eq. 2-28, Eq. 2-29, and Eq. 2-30 are inappropriately compounded, an upper limit of 2 is set on $M_{MAG\_Amount\ Released}$.

It is also expected that $M_{MAG\_Hole\ Diameter}$ operates in this manner over a limited range of event sizes; therefore, it is proposed that an upper limit of 3 and a lower limit of 0.3 be placed on this multiplier.

2.8.2.4   Material Being Released (MIE)

Section 2.3.4 discussed the effect of MIE on the probability of ignition in absolute terms. To create a PODI modifier, the earlier relationship needs to be converted to a relative term, and to do so a "typical" material MIE must be selected.

Hydrogen falls on the lower end of the MIE spectrum, with a value of approximately 0.015 mJ. At another extreme, ammonia has a MIE of 680 mJ. However, the most commonly used hydrocarbons (and the ones on which the data supporting the algorithms in Section 2.3.4 were based) fall in the range of 0.2–2 mJ. A baseline MEI value of 0.5 mJ was selected based on engineering judgment.

The original form of the relationship was in terms of an absolute PODI:

PODI = [0.15 – 0.25 log(MIE/0.5)], which can be rearranged to yield

$$PODI = 0.075 - 0.25 \log(MIE) \tag{2-33}$$

In the form of a PODI modifier, this relationship is more conveniently expressed as

$$M_{MAT} = 0.5 - 1.7 \log(MIE) \tag{2-34}$$

An upper limit of 3 is set for this multiplier to prevent it from having an overstated influence on the PODI. A lower limit of 0.1 is recommended on the basis that no matter how difficult a material is to ignite, there may still be some ignition source available that is strong enough to ignite it. These limits have the effect of treating all materials as having MIEs between 0.034 and 1.7 mJ. Therefore, if the analyst comes across a material that has an extremely high MIE and no correspondingly strong ignition source, the potential for ignition should be ignored.

2.8.2.5   Release/Flash Point/Boiling Point Temperatures

Section 2.3.5 proposed a relationship between ignition probability, the release temperature, and its boiling or flash point. Although this model is not intended to incorporate a dispersion model *per se*, a series of modeling runs were performed which indicated a linear relationship between the minimum distance to LFL and the difference between the normal boiling point (NBP) and release temperature of the material that might inform this algorithm; that is,

$$\text{Distance to LFL } \alpha \text{ (NBP} - T) \tag{2-35}$$

where T is the process temperature and NBP is the normal boiling point of the material (for mixtures, using the lower 10% boiling point). For liquid releases, the maximum distance was achieved when NBP and T were the same, and the dispersion distance dropped to near 0 when NBP – T was about 230 °F.

This corrective factor can be applied in a relative manner to vapor releases of the same material. This modifier can be applied for liquid releases as follows:

$$M_T = 1 - (NBP - T)/230 \qquad \text{(2-36)}$$

where M_T is the temperature modifier, NBP is the normal boiling point of the released material, and T is the normal process temperature (in degrees Fahrenheit).

Perhaps a more directly related physical property is flash point, since this defines whether a release generates enough vapors to be ignitable. The temperature modifier can alternatively be stated in flash point terms as follows:

$$M_T = 0.4 + (T - 1.3 \times FP)/230 \qquad \text{(2-37)}$$

In this case, FP is the flash point in degrees Fahrenheit.

These equations have a maximum value of 1. A minimum value of 0.001 is proposed to account for the possibility that the release might come into contact with an ignition source that is so warm that it could heat the material or otherwise provide ignitable conditions where they would not ordinarily exist.

2.8.2.6   Indoor Release

It is assumed that there is a somewhat increased chance of ignition indoors compared to outdoors, all else being equal, due to the lower dispersion of the flammable cloud. For this reason a multiplier of 1.5 suggested by subject matter experts is applied for indoor releases ($M_{IN/OUT}$). If the release is outdoors, $M_{IN/OUT} = 1$.

2.8.2.7   Combined Level 2 Algorithm for PODI

Incorporating each of the above contributors to PODI results in the following relationship:

$$PODI_{Level\ 2} = PODI_{S/D} \times M_{MAG} \times M_{MAT} \times M_T \times M_{IN/OUT} \qquad \text{(2-38)}$$

Of course, in principle this equation has a maximum value of 1. However, in the software tool it is limited to a maximum value of 0.9 so that a user does not inadvertently eliminate consideration of toxic or environmental outcomes in the event tree described at the start of the chapter.

2.9   ADVANCED (LEVEL 3) PROBABILITY OF IGNITION ALGORITHMS

2.9.1   Level 3 Algorithm for Probability of Immediate Ignition

An enhancement to the Level 2 algorithm for POII is simply to account for the temperature at the outlet of the point of discharge rather than using the temperature in the process. The difference may or may not be significant, depending on the circumstances of the release. However, this input requires an enthalpy calculation that is typically available in a consequence modeling package. Since this involves the results of another calculation, it is deemed too much effort for a Level 2 analysis and is relegated to Level 3.

In this refinement, the Level 3 algorithms are identical to those in Level 2, except that they use the revised temperature input. For the same reasons as described for a Level 1 analysis, the POII at Level 3 is limited to a maximum value of 0.99.

2.9.2   Level 3 Algorithm for Probability of Delayed Ignition

With some of the relevant parameters, no further Level 3 refinement is possible compared to what was used in Level 2. In other cases, some refinement is possible.

2.9.2.1   Strength/Duration of Ignition Sources

The strength and duration of the ignition sources are treated just as they are in the Level 2 analytical procedure described in Section 2.8.2.2, with one exception. The "strength" of an ignition source can be related to how well the ignition sources are controlled. Assuming the analyst has developed an unbiased tool for describing this level of control over ignition sources, the following multipliers can be applied to the value of S used in Eq. 2-28:

Multiply S by 0.7 for "good" control of ignition sources (the highest level of control that is practical).

Multiply S by 1.0 for "typical" control (the level of control that is considered to be the most realistic for a typical hazardous installation).

Multiply S by 1.5 for "minimal" control (the minimum level of control that may occur in practice), with the constraint that the resulting value of S cannot exceed 1.

Note that the factors above are not simply reflections of the level of controls exerted over an area, but rather the level of control relative to the industry norm. Thus the fact that roadways are typically nonclassified areas should not be penalized through these multipliers; that is taken into account in the "strength" factor of ignition sources for roadways. "Minimal" control in this context might be a situation where heavy vehicles were permitted to idle continuously in the roadway.

2.9.2.2   Release Magnitude

The release magnitude is treated just as it is in the Level 2 analysis described in Section 2.8.2.2.

2.9.2.3   Material Being Released

The material being released is treated just as it is in the Level 2 analysis described in Section 2.8.2.4. The modified inputs for MIE that are developed for POII are not used in this case under the assumption that delayed ignition will take place with vapor and not spray, and the temperature at the time of a delayed ignition will be similar to that used to develop MIE data (as opposed to being hot from the process or cooled due to Joule-Thompson effects) due to mixing with ambient air.

2.9.2.4   Indoor Events

Relative to Level 2, improvements can be proposed to account for either (a) advanced gas detection and ventilation systems in dispersing/removing flammable clouds from the building or (b) the greater chance for human ignition of flammable clouds relative to other ignition sources compared to an outdoor release. These two influences can be handled separately or together as described next.

Human Influence—This is described in Section 2.3.8.3 and will be implemented as follows—Assuming that the MIE is sufficiently low to allow human static ignition, the PODI for indoor releases will be related to the humidity in the room assuming that the average indoor process operation has a relative humidity of 50%. Based on the relationship of ignition to relative humidity in Section 2.3.8.3, this modifier is then

$$M_{RH} = [100 - 1.5 \times \text{relative humidity}] / 25 \qquad \textbf{(2-39)}$$

Where M_{RH} has a lower limit of 0. However, this multiplier applies only to the proportion of indoor events that are assumed to be ignited by people, which is assumed to be 20% in a Class 1/Division 2 location and 50% in a Class 1/Division 1 (or equivalent) location. The remainder of the ignitions are modified using the "generic" indoor multiplier or the detection/ventilation-influenced algorithms.

Detection and Ventilation—The method for evaluating the effect of detection and ventilation systems is described in Appendix B and is not repeated here. The resulting multiplier can be used independently of the "human influence" modifier below.

2.9.2.5   Mitigation Measures

In Level 3, the user is allowed to enter the benefits of measures that can be shown would lower the probability of ignition. One example of this is an isolation system that is used to limit the duration of an event; however, the benefit of this is best quantified in the strength/duration algorithms. Other approaches include deluge or inerting systems that activate upon detection of hydrocarbons. It is beyond the scope of this book to quantify the effectiveness and availability of such specialized systems for the infinite combinations of applications to which they might be applied. Therefore the user must define the probability of failure and enter this value (FIP). Do not double-count such measures by taking an FIP credit as well as the credit in Section 2.9.2.1 for "good ignition controls," however.

2.9.2.6   Combined Level 3 Algorithm for PODI

The combined Level 3 algorithm for PODI is then

$$PODI = PODI_{S/D} \times M_{MAG} \times M_{MAT} \times M_{IN/OUT} \times FIP \qquad \textbf{(2-40)}$$

Naturally, this equation has a lower limit of 0 and an upper limit of 1.

2.10   DEVELOPING INPUTS WHEN CHEMICAL PROPERTIES ARE NOT AVAILABLE

In some cases, important inputs such as MIE or AIT are not available in the literature. In such cases, surrogate estimates are needed.

In other cases, the necessary properties are available for pure chemicals but not for a mixture of those chemicals. In these cases, appropriate mixing rules are required.

2.10.1   Estimating Input Properties of Chemicals Not in the Pick List

In general, it will be necessary to estimate AIT and MIE for some chemicals either through other sources or by comparison to similar chemicals for which the AIT and/or MIE are known. However, estimating the AIT or MIE by comparison to other chemicals can be difficult.

2.10.1.1   AIT

As noted in Chapter 1, AIT cannot be correlated with more easily obtained parameters such as boiling point. A relationship between AIT and the chemical structure has been demonstrated for some hydrocarbon types which relates to the chain length and degree of branching (see Figure 1.3), but this is not generally useful for the full range of chemicals that might be considered by readers of this book.

In the absence of a generally applicable approach to AIT estimation, it is best to develop an estimate based on similar chemicals for which the AIT is known. There are several sources of this information, some of which is captured in Appendix A. However, there are multiple attributes of a chemical which should be considered, including the molecular weight, degree of branching, presence of "side" atoms such as chlorine, and more.

In general, it is preferred to use a "best estimate" rather than attempt to choose a "conservative" value since it may not be clear what the conservative approach would be. If one chooses a low AIT to be conservative, this results in a greater POII but a lower PODI. Thus, the analyst is suggesting that a fire is more likely than a delayed explosion, which may be either conservative or nonconservative, depending on the circumstances. Similarly, a high ignition probability may be associated with a low probability of a toxic outcome for releases of chemicals that are both flammable and toxic. So forcing the model to generate high ignition probabilities may be either conservative or nonconservative.

2.10.1.2   MIE

Some generalizations can be made with regard to MIE. Most paraffins with eight carbons or less have MIE values around 0.25 mJ, although the replacement of hydrogen atoms with other atoms can have dramatic effects. For example, methane (CH_4) has a MIE of 0.3 mJ while carbon disulfide (CS_2) has a much lower MIE of 0.015 mJ and methylene chloride (CH_2Cl_2) has a MIE of 10,000 mJ. Britton (2002) has developed a relationship between MIE and the heat of combustion of the fuel per mole of oxygen consumed (Figure 2.3).

$Y = M0 + M1 \cdot x + ... M8 \cdot x^8 + M9 \cdot x$	
M0	4.0056
M1	-0.06231
M2	0.00024333
R	0.98236

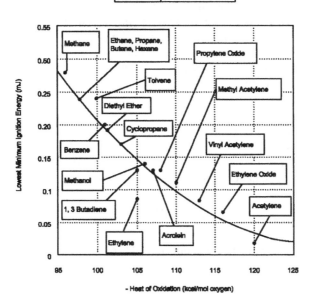

Figure 2.3. Dependence of MIE on heat of oxidation.

The relationship described in Figure 2.3 works best for hydrocarbons and related oxygenated compounds (CHO-type molecules). Some types of compounds were excluded (halogenated compounds, for example). Other compounds suppress the ignitability of flammables in mixtures. An example common to refining applications is that of HF- and LPG-range materials; a 25–30% concentration of HF is said to be sufficient to lower the ignition probability of the mixture to near zero.

Therefore, if the probability of ignition results are sensitive to the value of MIE used and the cost implications of the POII and PODI values are significant, it may be worthwhile to conduct a laboratory test to measure a value rather than depend on surrogates.

2.10.2 Estimating the Properties of Flammable Mixtures

There are two general classes of cases to consider when evaluating the effective properties of mixtures for the purpose of utilizing the tools in this book. In the first case, the mixture is homogeneous in the sense that it does not tend to separate into different phases upon release. A homogeneous mixture is assumed in the discussions of specific variables such as AIT in the remainder of this section.

A nonhomogenous mixture, as used in this book, is one in which the different components of the mixture separate shortly after release. A mixture of hydrogen and heavy hydrocarbons (as in a diesel oil hydrotreater) is an example of this. In the case of nonhomogeneous mixtures the two phases should be treated separately and the results combined thoughtfully. In the hydrotreater example, the hydrogen portion of the release may find an ignition source and burn back to the pool of diesel, igniting it as well. But depending on the release conditions and location, the hydrogen may float away without contacting an ignition source. Thus modeling this situation as a homogenous mixture may over- or underpredict the actual ignition probabilities.

2.10.2.1 AIT

The AIT of a mixture of fuels has been the subject of limited study. An example by Britton (1990b) involves mixtures of ethylene oxide. Figure 2.4 shows a comparison of experimentally developed mixture AITs vs. what would be predicted using LeChatalier's mixing rule:

$$AIT_{mix} = 1/[\Sigma x_i/AIT_i] \qquad (2\text{-}41)$$

where x_i and AIT_i are the mole fraction and AIT, respectively, of component i.

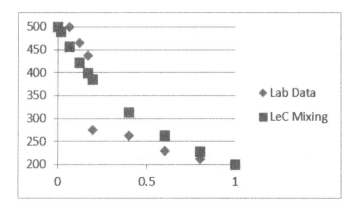

**Figure 2.4. Comparison of laboratory and LeChatalier
mixing rule values for AIT.**

The comparison is not perfect, but there is no existing rule that can universally address the inflection area of the data for all chemicals. As a result, LeChatalier's mixing rule is proposed for this purpose. Regardless of the assumed efficacy of mixing rules, where practical it is always preferred to perform actual laboratory tests of mixture flammable properties rather than depend on correlations.

2.10.2.2 MIE

The flammable properties of chemicals were explored at length in a paper by Britton (2002). Britton did not propose a mixing rule for MIE per se, but his findings may form the basis for a mixing rule. Using the algorithms developed by Britton for MIE, a very limited number of mixtures were checked against both LeChatalier ($MIE_{mix} = 1/[\Sigma x_i/MIE_i]$) and standard proportional behavior, that is,

$$MIE_{mix} = \Sigma\,[x_i \times MIE_i] \qquad\qquad \textbf{(2-42)}$$

In these checks, neither LeChatalier's rule nor the proportional mixing rules were found to be completely satisfactory, but an average of the two did provide a good prediction. It is not known if this conclusion can be generalized. Therefore, if the POII and PODI values are sensitive to the MIE value, laboratory testing may be appropriate.

Important too are mixtures that include diluents or suppressants to ignition activity, as mentioned in Section 2.10.1.2. It is beyond the scope of this book to address all the possibilities, and so the user should have such mixtures tested if necessary.

2.10.2.3 Reactivity

The propensity of a mixture to ignite explosively is, in part, a function of its fundamental burning velocity, referred to here as its "reactivity." Investigators have found some mixtures that obey LeChatalier's mixing rule for flame velocity, and others that follow standard proportional behavior, that is,

$$FFV_{mix} = \Sigma x_i \times FFV_i$$

(2-43)

If data for the particular mixture being evaluated is not available in the literature and if the implications of the method used are significant, it may be appropriate to have a test performed.

2.11 WORKED EXAMPLE

2.11.1 Problem Statement

As an illustration of the proposed algorithms, consider this example:

- Chemical released—Propane
- Phase of release—Vapor
- Process temperature—100 °F
- Process pressure—200 psig
- Location—Outdoor
- Hole diameter—2 inches
- Release duration—2 minutes
- Ignition source type—Process unit
- Fraction of source type that is in the cloud—0.20
- Ignition source control—Typical
- Reliability of detection/deluge/suppression system that prevents ignition of flammable releases—0.70

Not all of these inputs will be required for all levels of the analysis. Other inputs are left unstated because propane is a chemical with well-known properties.

2.11.2 Level 1 Analysis

In Level 1, only the following inputs are needed:

- Chemical released (propane)
- Process temperature (100 °F)
- Operation indoor or outdoor (outdoor)

The chemical defines the variable that is incorporated in Level 1—Propane has an AIT of 932 °F. This would be entered automatically from a pick list in the software version of these algorithms.

2.11.2.1 Level 1 POII

As per Section 2.7.1.2, since T/AIT < 0.9, then $P_{ai} = 0$ (that is, there is no chance that the release will autoignite). Then, per Section 2.7.1.3,

$$POII_{Level\ 1} = 0.05 + (1 - 0.05) \times P_{ai}, \text{ or}$$

$$POII_{Level\ 1} = 0.05 + (1 - 0.05) \times 0 = 0.05$$

2.11.2.2 Level 1 PODI

As per Section 2.7.2, the PODI is set at 0.25 for an outdoor release.

2.11.3 Level 2 Analysis

In Level 2, the following additional inputs are needed:

- Phase of release (vapor)
- Process pressure (100 psig)
- Hole diameter (2 inches)
- Release duration (2 minutes)
- Ignition source type—Process unit (20% coverage)

The chemical selected will define the "reactivity" (medium) and MIE (0.26 mJ), which are also inputs to Level 2.

2.11.3.1 Level 2 POII

As per Section 2.8.1.1, the POII for Level 2 is calculated as follows:

$$POII_{static} = 0.003 \times P^{1/3} \times MIE^{-0.6}$$

where P is in units of psig and MIE in units of millijoules. Then,

$$POII_{static} = 0.003 \times 100^{1/3} \times 0.26^{-0.6} = 0.03$$

The calculation for P_{ai} is the same as for Level 1. Then the overall POII for Level 2 is

$$POII_{Level\ 2} = P_{ai} + (1 - P_{ai}) \times POII_{static}, \text{ or}$$

$$POII_{Level\ 2} = 0 + (1 - 0) \times 0.03 = 0.03$$

2.11.3.2   Level 2 PODI

The Level 2 PODI includes a number of terms:

$$PODI_{Level\ 2} = PODI_{S/D} \times M_{MAG} \times M_{MAT} \times M_T \times M_{IN/OUT}$$

PODI$_{S/D}$—PODI$_{S/D}$ is related to the ignition source strength, and the duration of exposure, as per the following relationship from Section 2.8.2.2:

$$PODI_{S/D} = 1 - [(1 - S^2) \times e^{-St}]$$

where S is defined by the ignition source type that is input, as taken from Table 2.2 (0.2). The duration t is an input to this level (2 minutes). Then

$$PODI_{S/D} = 1 - [(1 - 0.2^2) \times e^{-0.2(2)}] = 0.36$$

M$_{MAG}$—As per Section 2.8.2.3, M_{MAG} is based on the magnitude of the release, expressed here as a function of the hole size. For vapors, M_{MAG} is equal to the hole diameter as measured in inches ($M_{MAG} = 2$).

M$_{MAT}$—M_{MAT} is a function of the MIE of the material being released as per Section 2.8.2.4:

$$M_{MAT} = 0.075 - 0.25 \log(MIE)$$

In this case, $M_{MAT} = 0.075 - 0.25 \log(0.26) = 0.22$

M$_T$—M_T is a temperature modifier related to material volatility and more as described in Section 2.3.5. For vapors, this has a value of 1.

M$_{IN/OUT}$—$M_{IN/OUT}$ is the multiplier used if a release is indoors. For this outdoor release, the value is 1.

Thus, the overall Level 2 PODI is

$$PODI_{Level\ 2} = 0.36 \times 2 \times 0.22 \times 1 \times 1 = 0.16$$

2.11.4   Level 3 Analysis

In Level 3, the user has the option of introducing both new terms and some replacement terms.

2.11.4.1   Level 3 POII

The user has the option of entering the postrelease temperature to replace the process temperature if the user has access to an appropriate discharge model. In this example it will be assumed that there is no discharge model readily available, and the postrelease temperature is assumed to be the same as the process temperature. The algorithms for POII remain the same as for Level 2 POII, and so the POII value is the same as before, 0.03.

2.11.4.2 Level 3 PODI

Level 3 PODI is treated the same as Level 2 for outdoor releases, with two exceptions.

The user has the opportunity to enter a value that describes how well ignition sources are controlled, which acts as a multiplier on the "S" term calculated in Level 2 $PODI_{S/D}$. In this case, the controls are "typical" and so the calculation is the same as for Level 2.

The user can also enter the impact of measures that would prevent ignition, if the reliability and effectiveness of the measures can be assessed properly. In this case a detection/deluge system has been described as being 70% effective in preventing ignition of flammable releases. Thus, the probability of failure of that detection/ deluge system (FIP) is equal to $1.0 - 0.70 = 0.30$.

The equation for Level 3 PODI is

$$PODI = PODI_{S/D} \times M_{MAG} \times M_{MAT} \times M_{IN/OUT} \times FIP$$
$$PODI = 0.36 \times 2 \times 0.22 \times 1 \times 1 \times 0.3 = 0.048$$

2.12 APPLICATION OF THE MODELS TO A STUDY WITH MULTIPLE IGNITION SOURCES

In some cases, the ignition sources for a given release will be dominated by a single ignition source, or perhaps the analysis is being conducted to measure the benefit of mitigating a specific release source. In these cases, the models described in this chapter will suffice.

In more complicated (and possibly more typical) situations, multiple ignition sources may be present. In most cases it is expected that the algorithms in this chapter can be used as is, with the ignition source strength being defined by an "area source" which by definition incorporates all the ignition sources in the space in which the flammable cloud is dispersed.

There will be situations in which an area-source approach is not sufficient. This may occur either because the approach is not a good physical representation of the area into which the cloud is being dispersed or because there are ignition sources besides those that are contained within the "area." In these cases, the algorithms described in this chapter need to be repeated and combined to obtain the best result.

The approach is to calculate the probability of immediate ignition, then calculate the probability of delayed ignition for each of the individual ignition sources following the methods in this chapter. The final probability of delayed ignition will be 1 minus the product of the probabilities of nonignition as follows:

$$PODI = 1 - \prod(1 - PODIi) \qquad (2\text{-}44)$$

where PODIi is the PODI of the ith ignition source as calculated using the methods in this chapter. As an example, the PODI for three ignition sources each having a probability of delayed ignition of 0.1, 0.2, and 0.5 is

$$PODI = 1 - (1 - 0.1)(1 - 0.2)(1 - 0.5) = 0.64$$

3 TECHNICAL BACKGROUND AND DATA SOURCES

3.1 INTRODUCTION AND SUMMARY

Table 3.1 provides a summary of information derived from other ignition data sources. The details of this information are provided in subsequent sections in this chapter.

Not all of these references are utilized in the algorithms in Chapter 2, but they are included for one or more of the following reasons:

- relevance to applications that are out of scope yet are still of interest to many readers of this book (e.g., offshore),
- applicability to circumstances that may be uncommon enough that they cannot be accounted for in the model yet would be useful to people who have that application, or
- historical significance.

To the extent that when gaps in the data are encountered or when data does not quite extend into the range of interest for a particular analysis, there is a temptation to extrapolate the available data. There are also some variables, or variable ranges, for which data cannot be practically collected. This chapter includes a significant amount of information based on "expert opinions." The users of this data and the developers of the algorithms in Chapter 2 must keep in mind the uncertainties inherent in the gaps/absence of data.

Table 3.1. Summary of ignition data references

Source	Variables Discussed	Relevance	Sect.
Spencer and Rew (1997)	Lists a wide variety of relevant variables. The following two papers expand on the concepts presented here.	Designed for onshore industry. Mainly references other sources.	3.2.1
Spencer, Daycock, and Rew (1998)	Includes different types of ignition source "strength", and introduces 1- $a^{-\lambda t}$ relationship between ignition probability and event duration.	Designed for onshore industry. Model based only indirectly on data.	3.2.1
Daycock and Rew (2004)	Expands on the Spencer and Rew work, proposing benefit factors for various types of ignition controls.	Designed for onshore industry. Benefits appear to be based on expert judgment and not "data" per se.	3.2.1
TNO Purple Book (2005)	Specifies POIs to use for Dutch risk assessments. Variables include flammability (flash point) and reactivity of released material and release rate (continuous releases) or quantity (instantaneous releases). A table of ignition source "strengths" is also given, similarly to Rew et al.	Onshore industry; also includes transport methods. Data + expert judgment. Should be considered for calibration purposes. Not detailed enough for the most sophisticated approaches in this book.	3.2.2
Cross-thwaite et al. (1998)	Early HSE risk assessment model, based on hole size. Also considers meteorology and release orientation.	Designed for LPG assessments. Model basis rather than directly "data." Specific to LPG and dated.	3.2.3
Thyer (2005)	Offshore ignition data, segregated by type/phase of release, electrical classification of area into which the material was released, and release size.	Offshore. Appears to be one of the better sources of true data. Not directly applicable to onshore operations but can use to calibrate algorithms to some degree and is the main source of data regarding the effect of electrical classification area.	3.2.4
Gummer and Hawksworth (2008)	Describes the possible mechanisms for spontaneous hydrogen ignition. Cites data that is assumed to be high immediate POI, low delayed POI. Also includes past incidents.	Hydrogen only. Somewhat vague and, relative to another source, possibly contradictory ignition data. Mainly of benefit for qualitative descriptors.	3.2.5 and 3.7.3
Cawley/U.S. Bureau of Mines (1988)	Is the only paper found that treats extremely low ignition probabilities, in this case caused by various electrical circuit strengths.	Source is mining industry. Data limited to a single mixture (8.3% methane in air) but is unique in providing actual data for low-energy sources.	3.2.6
HMSO/ Canvey (1981)	Early LPG risk assessment model.	Specific to LPG and is very dated. Obsolete for the purposes of this effort.	3.2.7
Witcofski (1981)	Discusses dispersion of releases of large volumes of liquid hydrogen.	Notable absence of ignition in all cases, relevant to discussions of hydrogen, low temperature, liquid phase releases, etc.	3.2.8

Table 3.1 (continued)

Source	Variables Discussed	Relevance	Sect.
Cox, Lees, and Ang (1990)	Primarily relates POI to the phase and release rate of material.	Seminal work in the POI field. A bit outdated at this point and based on offshore data, but still widely cited and used both offshore and onshore.	3.3.1
E&P Forum (1996)	Quotes several of the other sources mentioned here. New insight is provided on the relative strength of various types of ignition sources, in terms of the fraction of offshore ignitions caused by each type of ignition source.	Offshore based, but good data. No direct correlation between the ignition source data and the algorithms developed here, since the numbers of sources are not known. But it does provide a point of reference.	3.3.2
API (2000)	Considers the chemical type, phase, and whether it is above/below AIT.	Oil refining based. Based on reasonable judgment of authors. Concise; something like this might be a suitable framework for a "Level 1" algorithm. Does not include at least one significant variable (release rate) or provide new insights.	3.3.3
API (2003)	Evaluates potential for ignition by hot surfaces. Considers temperature of the hot surface as well as exposure duration and wind speed.	Oil industry based. Data limited, but very good in exposing the significance of hot-surface ignition variables. Provides new insights on the subject of hot-surface ignition.	3.3.4
Hamer et al. (1999)	Effect of hot surfaces (specifically induction motors) on ignition.	Oil industry based. Provides support for API 2003 work and relates it to a real-life hot-surface situation.	3.3.5
UKEI (2006)	Sophisticated model incorporating many variables. Developed for overlaying onto a plant grid.	Main focus on LPG and offshore facilities, but onshore also included. Very good on spatial variables; less detailed in chemical property-related variables. Useful for the same reasons that the Rew et al. work is useful.	3.3.6
Ronza et al. (2007)	Considers the material released and the amount released.	Very relevant for liquid transport. Based on real field data, albeit for non-"process" situations where the ignition source profile is likely to be different. Provides useful corroboration on the importance of liquid type and release amount.	3.4.1
Foster and Andrews (1999)	Considers many variables relevant to the probability of explosion.	Offshore-based. Primarily a "model" and not a release of new data. Model is too complicated to be incorporated into a general-purpose tool developed in this book.	3.4.2

Table 3.1 (continued)

Source	Variables Discussed	Relevance	Sect.
Srekl and Golob (2009)	Evaluates building fires, mainly non-industrial. Considers amount of flammables available, number of ignition sources available, and number of hours per day the building is active.	Good data, but limited in scope and applicability to process situations. Introduces at least one variable (fraction active time) that might be considered for indoor event POIs. Discounts another (amount of flammables present).	3.4.3
Duarte et al. (1998)	Reviews POI from hot surfaces and the influence of vapor flow around the hot surface.	In general terms, this source confirms findings of other investigators.	3.4.4
Swain et al. (2007)	Reviews POI of hydrogen releases, particularly with respect to release velocity.	Provides insights to cases where higher flow rates could result in lower POIs.	3.4.5
Dryer et al. (2007)	Evaluates hydrogen POIs, particularly with respect to source pressure and equipment configuration at the point of release	Good, lab based. Limited with respect to experimental configuration used. Similar to Swain insights and also suggests importance of the geometry at the point of release.	3.4.6
Britton (1990a)	Considers release pressure/velocity, temperature, orifice size.	Data very good for the limited chemicals considered. Provides insights into non-"standard" chemical classes, some of which are relevant to standard chemical types.	3.4.7
Pesce et al. (2012)	Builds on the Energy Institute model	Incorporates chemical properties, as defined by electrical classification groups	3.4.8
Spouge (1999)	Relates POI to release phase, release rate, and event duration.	Offshore oil/gas production. Mostly based on incident data; duration method based on judgment. Broad-based POI approach similar to API RBI and others.	3.5.1
Moosemiller (2010)	Evaluates POI and probability of explosion taking into account several source and general environmental inputs. Proposes method for indoor releases.	Onshore process industry. Some methods data based, some based on expert judgment. Covers more variables than most. Similar in scope and structure to the effort in this book, although with somewhat less detail.	3.5.2
Johnson (1980)	Considers humans as potential ignition sources.	Data useful but limited in scope. Suggests that in most cases human ignition sources are negligible in an industrial setting (with possible exception of indoor releases).	3.5.3
Jallais (2010)	Compares hydrogen POI models from several sources.	Hydrogen POI ignition model comparison only. Mainly illustrates inconsistencies of current methods	3.5.4

Table 3.1 (continued)

Source	Variables Discussed	Relevance	Sect.
Zalosh et al. (1978)	Analysis of reported hydrogen event outcomes, distributed by sources of release and ignition source.	Actual data, with unknown reporting bias. Useful as a point of calibration in a general sense. Since data sources are not well defined, it is difficult to draw precise conclusions from this data.	3.5.5
Smith (2011)	Comparison of petroleum to natural gas pipeline ignition probabilities.	Describes effects of static in the larger POIs observed in gas pipelines.	3.5.6
Lee et al. (1996)	Ignition of multicomponent liquid sprays. Relates the MIE of the liquids to the droplet diameter and suggests that the LeChatalier mixing rule may be appropriate for estimating a composite mixture MIE.	Lab data for liquid sprays (illustrated with heptane and decane). Useful insight of relevance of droplet size to ignition probability of liquid sprays, although the relevance to large-scale releases is not clear.	3.6.1
Babrauskas (2003)	This book is an encyclopaedic treatment of the subject of ignitions. Like Lee et al., this reports on the effect of spray droplet diameter on MIE, but for a wider variety of chemicals. There is also a table that notes the relationship between MIE and the temperature of the spray.	Same as Lee et al., plus provides MIE-temperature relationship.	3.6.2
Britton (1999)	Includes case histories. There is no data provided, but insights to various ignition factors are illustrated.	Provides useful insights to unusual ignition situations that might be missed in a review of data alone.	3.7.1
Pratt (2000)	Discusses mechanisms for offshore ignitions.	Provides insights to ignition mechanisms for offshore releases, which might be used as a basis for considering how to incorporate the more extensive offshore data into a tool for onshore studies. Expert judgment based.	3.7.2

There are some examples of extrapolation of previously developed relationships that will clearly be nonsensical (e.g., if they result in POIs that are less than 0 or greater than 1). Other extrapolations are more subtle, for example, where there are limitations to the physical phenomena that control the variable. Thus there will always be need for user judgment in using this data and the tools that are developed from them.

3.2 GOVERNMENT-DRIVEN STUDIES

3.2.1 Rew et al.

The group including P. J. Rew, H. Spencer, and J. Daycock have probably been the most prolific publishers in the field of ignition probabilities, creating a series of reports for the U.K. Health and Safety Executive in the late 1990s and early 2000s. These reports are closely related to each other and so are presented here collectively.

3.2.1.1 Ignition Probability of Flammable Gases (CRR 146/1997)

In this report (Spencer and Rew, 1997, p. 2), the authors identify the following relevant parameters:

Nature of ignition source:
- *Continuous or intermittent;*
- *Strength;*
- *Design (for example, intrinsically safe);*
- *Type (for example, hot work, flare, electrical faults, static etc.);*
- *Location (onsite or offsite)*
- *Density of sources per unit area of land.*

Release location:
- *Enclosed or open;*
- *Distance to ignition sources (delayed or immediate ignition).*

Release type:
- *Fuel type (minimum ignition energy);*
- *Concentration of gas release (flammable limits, mean and intermittent);*
- *Self-generation of ignition (for example, static or mechanical sparks).*

There follows a discussion of each of these factors, along with some limited data and some other expert opinion on how these factors contribute to the ignition probabilities. Models that have been developed by others for some of these variables are listed, along with a conclusion that there are significant differences in the predictions made by the models.

Spencer and Rew (1997) proposed a model framework in which the "ignition probability is calculated by considering whether the flammable gas cloud will reach defined ignition sources within urban, rural or industrial locations, i.e. *it is based on site information rather than on historical data....*" [italics added here for emphasis]. Thus Spencer and Rew appear to acknowledge the difficulty in applying historical data to specific plant situations.

3.2.1.2   A Model for the Ignition Probability of Flammable Gases, Phase 2 (CRR 203/1998)

In this follow-up work (Spencer et al., 1998), the authors progress the development of POI predictions. They first propose the following equation for *non*-ignition Q_A:

$$Q_{A=\,Q_{A1}Q_{A2}\cdots\,Q_{AJ}} = \prod_{j=1}^{J} Q_{Aj} = \prod_{j=1}^{J} \{exp\,\{\mu_j A[(1 - a_j p_j)e^{-\lambda_j p_j t} - 1]\}\} \qquad \textbf{(3-1)}$$

where $\prod$ indicates the product of a series of terms. In this equation μ is the ignition source density, p is the probability of ignition from a source given that it is active and in the cloud, λ is the frequency that the ignition source is active, and a is the proportion of time that the ignition source is active.

This equation is then supported by a summary of offsite and industrial ignition source densities that was based on published statistics (Table 3.2 and Table 3.3). Of course, offsite ignition source densities in particular are likely to be different in developed vs. developing countries and will vary depending on land use planning practices.

There are a number of assumptions that were employed in the development of the tables from raw data such as certain types of facilities only operating during daytime hours. The proposed model may be useful within the context of a single user's system or for use by a user's group but is too complex for the purposes of this book. Therefore, the model itself is not adopted, although elements of the model such as ignition strength and time dependency of POI are utilized in the algorithms in Chapter 2.

Table 3.2. Summary of industrial ignition source data
(Spencer et al., 1998)

Source	Location	A_i	p_i	λ_i (per min)	a_i	μ_i (per hectare) Day	μ_i (per hectare) Night
Food Products	Indoor	15	0.25	0.056	0.99	0.097	0.015
	Outdoor	∞	1	0.0083	0.042	0.037	0.006
	Outdoor	∞	1	0.0083	0.281	0.059	0.009
Textiles	Indoor	15	0.15	0.056	0.99	0.163	0.016
	Outdoor	∞	1	0.0083	0.042	0.072	0.007
	Outdoor	∞	1	0.0083	0.281	0.091	0.009
Wood & Paper	Indoor	15	0.3	0.035	0.98	0.113	0.008
	Outdoor	∞	1	0.0083	0.042	0.053	0.004
	Outdoor	∞	1	0.0083	0.281	0.059	0.004
Printing	Indoor	15	0.8	0.0277	0.883	0.265	0.066
	Outdoor	∞	1		0.125	0.127	0.032
Chemicals	Indoor	15	0.6	0.023	0.99	0.117	0.020
	Outdoor	∞	1	0	1	0.018	0.003
	Outdoor	∞	1	0	0.25	0.062	0.011
Nonmetal	Outdoor	∞	1	0	1	0.062	0.021
Basic Metals	Outdoor	∞	1	0	1	0.028	0.009
Metal Products	Indoor	15	1	0.039	0.692	0.271	0.068
	Outdoor	∞	1	0	0.125	0.143	0.036
Machinery	Indoor	15	1	0.022	0.584	0.140	0.035
	Outdoor	∞	1	0	0.125	0.081	0.020
Electrical	Indoor	15	0.4	0.0347	0.98	0.145	0.014
	Outdoor	∞	1	0.0083	0.042	0.065	0.006
	Outdoor	∞	1	0.0083	0.2813	0.080	0.008
Transport	Indoor	15	1	0.022	0.584	0.051	0.013
	Outdoor	∞	1	0	0.125	0.029	0.007
Other	Indoor	15	0.6	0.037	0.862	0.170	0.026
	Outdoor	∞	1	0	0.25	0.077	0.012
Vehicle Repair	Outdoor	∞	0.4	0.042	0.861	0.115	0.000
Wholesalers	Indoor	15	0.3	0.0167	0.25	0.564	0.000
	Outdoor	∞	1	0.033	0.0033	0.564	0.000
Road Vehicles	Outdoor	∞	0.1	0	1	0.510	0.130
Trains	Outdoor	∞	0.5	0	1	0.000	0.000
Traffic Lights	Outdoor	∞	1	$0.1^d / 0.05^n$	0	0.004	0.004

^{d} = day, ^{n} = night

Table 3.3. Ignition sources in urban and rural areas during the day and night (Spencer et al., 1998)

Source	Location	A_i (ach)	Land Use	Time	μ_i (per hectare)	P_j	λ_i (per min)	a_j
Road Vehicles	Outdoor	∞	Urban	Day	0.51	0.1	0	1
				Night	0.13	0.1	0	1
			Rural	Day	0.027	0.1	0	1
				Night	0.0068	0.1	0	1
Traffic Lights	Outdoor	∞	Urban	Day	0.004	1	0.02-1	0
				Night	0.004	1	0-0.01	0
Trains	Outdoor	∞	Urban	Day	2.1×10^{-4}	0.5	0	1
				Night	7.4×10^{-5}	0.5	0	1
			Rural	Day	2.6×10^{-5}	0.5	0	1
				Night	9.2×10^{-6}	0.5	0	1
Balanced Flue Gas Appliances	Outdoor	∞	Urban	Day	2.33	1	0	0.05
				Night	2.33	1	0	0.125
			Rural	Day	1.7×10^{-3}	1	0	0.05
				Night	1.7×10^{-3}	1	0	0.125
Occasional Fires	Outdoor	∞	Urban	Day	8.28	1	2.2×10^{-5}	2.6×10^{-2}
				Night	8.28	1	3.4×10^{-6}	4.1×10^{-4}
			Rural	Day	0.20	1	2.5×10^{-4}	3.0×10^{-2}
				Night	0.20	1	5.7×10^{-6}	6.8×10^{-4}
Households	Indoor	2	Urban	Day	8.28	1	0	0.5
				Night	8.28	1	0	0.5
			Rural	Day	0.20	1	0	0.5
				Night	0.20	1	0	0.5
Restaurants and Public Houses	Indoor	2	Urban	Day	0.034	1	0	0.5
				Night	0.034	1	0	0.3
			Rural	Day	9×10^{-4}	1	0	0.5
				Night	9×10^{-4}	1	0	0.3
Shops	Indoor	2	Urban	Day	0.27	1	0	0.75
			Rural	Day	0.007	1	0	0.75
Hospitals	Indoor	2	Urban	Both	9×10^{-4}	1	0	1
Offices	Indoor	2	Urban	Day	0.16	1	0	0.75

3.2.1.3   Development of a Method for the Determination of Onsite Ignition Probabilities (RR 226)

This report (Daycock and Rew, 2004) compiles the previous work by this group and expands upon it. Early in the document there is a good discussion of the control of ignition sources, which, while not adopted as is in this book, covers the same topics.

The report goes on to suggest various ignition probability modification factors, including those in Table 3.4 through Table 3.8:

Ignition Source Controls: Comment on Use of "Systems" Modifiers—In general there is an assumption in this book that there is a base level of competency in the PSM systems of the facilities being analyzed. When applying event probability modifiers that are based on the level of PSM competence, how is consistency in the assessment maintained, and can the typical QRA specialist do this assessment? Is an audit on a facility performed before doing the QRA or LOPA to which this analysis will be applied?

There are pros and cons of using such modifiers. On the one hand, it seems only fair to "reward" excellence in PSM and indirectly to provide an economic incentive for such excellence. However, there are some logical pitfalls with that approach. For example, assume that these modifiers were used in a LOPA analysis and, as a result, the analysis suggested a higher integrity interlock than would have been the case without the modifier. If the plant is not maintaining the electrical systems (a symptom of the poor PSM that prompted the higher integrity interlock), does the "improved" interlock actually provide added reliability? These and similar questions should be considered in the use of PSM and other system modifiers.

**Table 3.4. Effectiveness of ignition source controls
(Daycock and Rew, 2004)**

Ignition Control	Factor		Overall Probability of Ignition
Ideal	0	None	Design and maintenance ensures no ignition source at any time
Excellent	0.1	Minimal	Well designed and maintained— ignition only arising from rare events
Typical (good)	0.25	Limited	Designed to meet standards and maintained regularly—ignition eliminated in normal operation, but potential for failure of systems or changing circumstances to result in occasional ignition source
Poor	0.5	Poor	Does not meet precise standards and poorly maintained— significant potential for ignition sources to occur
None	> 0.5	None	No adherence to standards and little maintenance—significant potential for ignition sources to occur

Ignition Source Parameters—In Table 3.5 through Table 3.8, the abbreviations/definitions are:

p = probability that will occur when the source is active and in contact with the gas

ta = period between activation of intermittent ignition sources

ti = period during which the source is active

a = probability that the source is active initially [= ta/(ta + ti)]

λ = frequency (per minute) at which the source becomes able to ignite the gas (= 1/(ta + ti)

As an example, consider a single release source case where the ignition source is described by three of the rows from Table 3.5 (as listed in Table 3.6), and the ignition source is 100% covered by the flammable cloud for 1 minute. Then, applying Eq. 3-1,

$$Q_{A= Q_{A1}Q_{A2\cdots} \; Q_{AJ}} = \prod_{j=1}^{J} Q_{Aj} = \prod_{j=1}^{J} \{exp \; \{\mu_j A[(1 - a_j p_j)e^{-\lambda_j p_j t} - 1]\}\}$$

results in the following:

μ = assumed density of ignition sources per hectare

Table 3.5. Ignition sources in urban and rural areas during the day and night (Daycock and Rew, 2004)

Ignition Source Parameters (for typical plant)

Land-use type	Ignition sources	Base case, or 'typical', ignition source parameters						
		p	ta	ti	a	λ	μ	Loc.
1. Car park	'Rush hour' vehicles	0.2	6	474	0.0125	0.0021	160	Out.
	'Other' vehicles	0.2	6	54	0.1	0.0167	3	Out.
	Smoking	1	10	470	0.021	0.0021	8	Out.
2. Road area	'Rush hour' vehicles	0.1	6	474	0.0125	0.0021	160	Out.
	'Other' vehicles	0.1	6	54	0.1	0.0167	3	Out.
	Delivery vehicles	0.1	6	24	0.2	0.0333	20	Out.
	Traffic control	1	0	15	0	0.0667	20	Out.
3. Controlled roads	Delivery vehicles	0.2	6	24	0.2	0.0333	20	Out.
4. Waste ground	None	0	-	-	0	0	0	Out.
5. Boiler house	Boiler	1	120	360	0.25	0.0021	200	In.
6–11. Flames	Continuous (indoors)	1	-	0	1	0	200	In.
	Continuous (outdoors)	1	-	0	1	0	200	Out.
	Infrequent (indoors)	1	60	420	0.125	0.0021	200	In.
	Infrequent (outdoors)	1	60	420	0.125	0.0021	200	Out.
	Intermittent (indoors)	1	5	55	0.0833	0.0167	200	In.
	Intermittent (outdoors)	1	5	55	0.0833	0.0167	200	Out.
12. Kitchen facilities	Smoking	1	5	115	0.042	0.0083	200	In.
	Cooking equipment	0.25	5	25	0.167	0.0333	100	In.
13-15. Process areas	'Heavy' equipment levels	0.5	-	-	1	0.028	50	In.
	'Medium' equipment	0.25	-	-	1	0.035	50	In.
	'Light' equipment levels	0.1	-	-	1	0.056	50	In.
16. Classified	None	0	-	-	0	0	0	In.
17. Classified (Ex.)	Material handling	0.05	5	25	0.167	0.0333	10	Out.
18. Storage (Ex.)	Material handling	0.1	10	20	0.333	0.0333	10	Out.
19. Office	'Light' equipment levels	0.05	-	-	1	0.056	20	In.

Table 3.6. Example calculations of Daycock and Rew algorithm

Ignition Source	p	a	λ	μ	A	t	Prob. Non-Ign	Prob. Ign.
Flame, Continuous	1	1	0	200	1	1	0	1
Car Park, Rush Hour	.2	.0125	.0021	160	1	1	.627	.373
Storage Area Material Handling	.1	.333	.0333	10	1	1	.694	.306

Table 3.7. Ignition source parameters for plants with "good" ignition controls (Daycock and Rew, 2004)

Land-use type	Ignition sources	Ignition source parameters with 'good' quality ignition controls						
		p	ta	ti	a	λ	μ	Loc.
1. Car park	'Rush hour' vehicles	0.2	6	474	0.0125	0.0021	160	Out.
	'Other' vehicles	0.2	6	54	0.1	0.0167	3	Out.
	Smoking	0	10	470	0.021	0.0021	8	Out.
2. Road area	'Rush hour' vehicles	0.1	6	474	0.0125	0.0021	160	Out.
	'Other' vehicles	0.1	6	54	0.1	0.0167	3	Out.
	Delivery vehicles	0.1	6	24	0.2	0.0333	20	Out.
	Traffic control	0	0	15	0	0.0667	20	Out.
3. Controlled roads	Delivery vehicles	0.2	6	24	0.2	0.0333	20	Out.
4. Waste ground	None	0	-	-	0	0	0	Out.
5. Boiler house	Boiler	0.5	120	360	0.25	0.0021	200	In.
6-11. Flames	Continuous (indoors)	0.5	-	0	1	0	200	In.
	Continuous (outdoors)	0.5	-	0	1	0	200	Out.
	Infrequent (indoors)	0.5	60	420	0.125	0.0021	200	In.
	Infrequent (outdoors)	0.5	60	420	0.125	0.0021	200	Out.
	Intermittent (indoors)	0.5	5	55	0.0833	0.0167	200	In.
	Intermittent (outdoors)	0.5	5	55	0.0833	0.0167	200	Out.
12. Kitchen facilities	Smoking	0	5	115	0.042	0.0083	200	In.
	Cooking equipment	0.1	5	25	0.167	0.0333	100	In.
13-15. Process areas	'Heavy' equipment levels	0.2	-	-	1	0.028	50	In.
	'Medium' equipment	0.1	-	-	1	0.035	50	In.
	'Light' equipment levels	0	-	-	1	0.056	50	In.
16. Classified	None	0	-	-	0	0	0	In.
17. Classified (Ex.)	Material handling	0.05	5	25	0.167	0.0333	10	Out.
18. Storage (Ex.)	Material handling	0.1	10	20	0.333	0.0333	10	Out.
19. Office	'Light' equipment levels	0.05	-	-	1	0.056	20	In.

Bold numbers in the table indicate parameters that are changed from the base case

Table 3.8. Ignition source parameters for plants with "poor" ignition controls (Daycock and Rew, 2004)

Land-use type	Ignition sources	Ignition source parameters with 'poor' quality ignition controls						
		p	ta	ti	a	λ	μ	Loc.
1. Car park	'Rush hour' vehicles	**0.3**	6	474	0.0125	0.0021	160	Out.
	'Other' vehicles	**0.3**	6	54	0.1	0.0167	3	Out.
	Smoking	1	10	470	**0.042**	0.0021	8	Out.
2. Road area	'Rush hour' vehicles	**0.2**	6	474	0.0125	0.0021	160	Out.
	'Other' vehicles	**0.2**	6	54	0.1	0.0167	3	Out.
	Delivery vehicles	**0.2**	6	24	0.2	0.0333	20	Out.
	Traffic control	1	0	15	0	**0.1333**	20	Out.
3. Controlled roads	Delivery vehicles	0.2	6	24	0.2	0.0333	20	Out.
4. Waste ground	None	0	-	-	0	0	0	Out.
5. Boiler house	Boiler	1	120	360	0.25	0.0021	200	In.
6–11. Flames	Continuous (indoors)	1	-	0	1	0	200	In.
	Continuous (outdoors)	1	-	0	1	0	200	Out.
	Infrequent (indoors)	1	60	420	0.125	0.0021	200	In.
	Infrequent (outdoors)	1	60	420	0.125	0.0021	200	Out.
	Intermittent (indoors)	1	5	55	0.0833	0.0167	200	In.
	Intermittent (outdoors)	1	5	55	0.0833	0.0167	200	Out.
12. Kitchen facilities	Smoking	1	5	115	0.042	0.0083	200	In.
	Cooking equipment	**0.5**	5	25	0.167	0.0333	100	In.
13-15. Process areas	'Heavy' equipment levels	**1**	-	-	1	0.028	50	In.
	'Medium' equipment	**0.5**	-	-	1	0.035	50	In.
	'Light' equipment levels	**0.2**	-	-	1	0.056	50	In.
16. Classified	**Material handling**	**0.05**	**5**	**25**	**0.167**	**0.0333**	**50**	**In.**
17. Classified (Ex.)	Material handling	**0.1**	5	25	**0.333**	0.0333	10	Out.
18. Storage (Ex.)	Material handling	0.1	10	20	**1**	**0**	10	Out.
19. Office	'Light' equipment levels	0.05	-	-	1	0.056	20	In.

Bold numbers in the table indicate parameters that are changed from the base case

The ignition model that is ultimately developed by Daycock and Rew is useful within the prescribed software package but cannot be as readily adapted as the more broad-based, "open-source" effort being developed in this book. One of the authors also noted that their method has not been validated for releases with very low (<0.01) probability of ignition. However, the general form of the ignition probability model illustrated above is a useful approach to adopt.

3.2.2   Bevi Risk Assessment Manual (TNO Purple Book)

In this work, well-known as the TNO Purple Book, the Dutch regulatory authors propose values and algorithms for use in quantitative risk assessments. However, the Purple Book has been superseded for the purposes of regulatory-driven risk assessments by the "Reference Manual Bevi Risk Assessments" (RIVM, 2009). The prescribed ignition probability values are provided in Table 3.9, Table 3.10, and Table 3.11.

It is not clear to what extent this information is based on actual "data" as opposed to expert opinion and conjecture. The fact that the inside vs. outside numbers are a round factor of 2 from each other suggests more the latter, although doubtless the authors considered the available data.

The term "direct ignition" appears to refer to an event that self-ignites, as opposed to reaching another ignition source such as a fired heater. An example cited in the Purple Book describes an instantaneous release in which "direct ignition" did not occur until a vapor cloud had formed in the vicinity of the release. Thus the term is not exactly the same as "immediate ignition" as used in this book but will often have the same net effect.

Table 3.9. Probability of direct ignition for stationary installations

Substance Category	Source Term Continuous	Source Term Instantaneous	Probability of Direct Ignition
Category 0 Average/High Reactivity	< 10 kg/s	< 1,000 kg	0.2
	10–100 kg/s	1,000–10,0000 kg	0.5
	> 100 kg/s	> 10,000 kg	0.7
Category 0 Low Reactivity	< 10 kg/s	< 1,000 kg	0.02
	10–100 kg/s	1,000–10,0000 kg	0.04
	> 100 kg/s	> 10,000 kg	0.09
Category 1	All flow rates	All quantities	0.065
Category 2	All flow rates	All quantities	0.01
Category 3,4	All flow rates	All quantities	0

Table 3.10. Probability of direct ignition of transport units in an establishment

Substance Category	Transport Unit	Scenario	Probability of Direct Ignition
Category 0	Road tanker	Continuous	0.1
	Road tanker	Instantaneous	0.4
	Tank Wagon	Continuous	0.1
	Tank Wagon	Instantaneous	0.8
	Ships—gas tankers	Continuous, 180 m^3	0.7
	Ships—gas tankers	Continuous, 90 m^3	0.5
	Ships—semi-gas tankers	Continuous	0.7
Category 1	Road tanker, tank wagon ships	Continuous, instantaneous	0.065
Category 2	Road tanker, tank wagon ships	Continuous, instantaneous	0.01
Category 3, 4	Road tanker, tank wagon ships	Continuous, instantaneous	0

Table 3.11. Classification of flammable substances

Category	WMS Category	NFPA Class	Limits
Category 0	Extremely flammable	IA	Liquid substances and preparations with a flash point lower than 0 °C and a boiling point (or the start of a boiling range) less than or equal to 35 °C.
			Gaseous substances and preparations that may ignite at normal temperature and pressure when exposed to air.
Category 1	Highly flammable	IB	Liquid substances and preparations with a flash point below 21 °C, which are not, however, extremely flammable.
Category 2	Flammable	IC/II	Liquid substances and preparations with a flash point greater than or equal to 21 °C and less than or equal to 55 °C.
Category 3		IIIA	Liquid substances and preparations with a flash point greater than 55 °C and less than or equal to 100 °C.
Category 4		IIIB	Liquid substances and preparations with a flash point greater than 100 °C.

Notes:
1. For loading scenarios, the probabilities of ignition are based on Table 3.2.
2. If the process temperature of Category 2, Category 3, and Category 4 substances is higher than the flash point, then the direct probability of ignition for Category 1 substances is used.
3. The NFPA class column entries represent the closest fit to the NFPA class types and are not exact comparisons.

Another table and equation clearly describe the POI as a function of ignition source "strength" and exposure duration; the assumed application here is for releases that occur or migrate offsite. Table 3.12 is a compilation of data taken from tables in "RIVM" and the Purple Book. The numbers below assume that the flammable cloud is in contact with the ignition source.

Table 3.12. Probability of ignition of a flammable cloud during a time window of one minute for a number of sources

Source Type	Ignition Source	Probability of Ignition
Point Source	Adjacent process installation	0.5
	Flare	1.0
	Oven (outside)	0.9
	Oven (inside)	0.45
	Boiler (outside)	0.45
	Boiler (inside)	0.23
	Motor vehicle	0.4
	Ship	0.5
	Ship transporting flammable materials	0.3
	Diesel train	0.4
	Electric train	0.8
Line Source	High-voltage cable (per 100 meters)	0.2
	Road/Railway	See Note below
Area Source	Chemical plant	0.9 per site
	Oil refinery	0.9 per site
	Heavy industry	0.7 per site
	Light industrial warehousing	Treat as population source
Population source	Households (per person)	0.01
	Offices (per person)	0.01

Note on Road/Railway Line Sources: The ignition probability for a road or railway near the establishment or transport route under consideration is determined by the average traffic density.

[Re: Table 3.12 note above] The average traffic density, d, is calculated as

$$d = NE/v \qquad \text{(3-2)}$$

Where:

N	=	number of vehicles per hour	(h^{-1})
E	=	length of a road or railway section	(km)
v	=	average velocity of vehicle	$(km\ h^{-1})$

If $d \leq 1$, the value of d is the probability that the source is present when the cloud passes; the probability of an ignition in the time interval 0 to t, $P(t)$, equals

$$P(t) = d(1 - e^{-\omega t}) \qquad \text{(3-3)}$$

where:

ω = the ignition effectiveness of a single vehicle (s^{-1})

If $d \leq 1$, the value of d is the average number of sources present when the cloud passes; the probability of an ignition in the time interval 0 to t, $P(t)$, equals

$$P(t) = d(1 - e^{-d\omega t}) \qquad \text{(3-4)}$$

where:

ω = the ignition effectiveness of a single vehicle (s^{-1})

As an illustration of the equations above, consider a lightly traveled plant road with the following parameters, as defined above:

N = 20 per hour

E = 0.03 km (assumed flammable cloud width as it contacts the road)

v = 25 km per hour

ω = 0.4 per minute ~ 0.007 per second

Assume further that there is a 5-minute timeframe before the road is isolated and/or the traffic realizes it should not be passing through the cloud.

Then

d = 20 × 0.03/25 = 0.024 and P = 0.024[1 − $e^{-(0.024)(0.007)(300)}$] = 0.0012

Commentary on Table: As is usually the case with existing ignition probability data, Table 3.12 is based primarily on expert opinion, and individual companies might have different experiences. One company, for example, has noted that high-voltage power lines have ignited gas clouds both times that a flammable cloud has reached the line, although this could be a function of a long-duration exposure, an easily ignitable material, or simply bad luck.

3.2.3   HSE/Crossthwaite et al.

These investigators considered ignition probabilities for risk assessments of LPG facilities (Crossthwaite et al., 1988). Figure 3.1 is an example of one of the event trees presented by the authors; others are presented for "limited vessel failures" and "leaks from piping" but are not reproduced here. Note that certain values (e.g., fraction of time at F2 or D5 meteorology) were fixed in a software program to ensure consistency.

This shows the event tree form, incorporating conditional probabilities such as release orientation and meteorology, as well as ignition probabilities. The information is somewhat dated, however, and so is used only as a point of reference in this work.

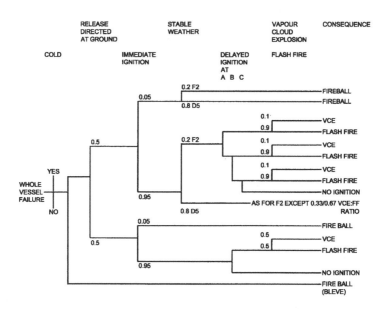

Figure 3.1. Ignition probability for whole vessel failures.

3.2.4   HSE/Thyer

Table 3.13 is a review of offshore data sponsored by the HSE (Thyer, 2005) that shows the probabilities of ignition for various release subsets, including the effect of release size, phase, and area classification.

Table 3.13. Probabilities of ignition for various
release subsets (Thyer, 2005)

Fluid Type	Area Classification	Release Size	Number of Releases	Number Ignited	% Ignited	Approximate Probability
Oil	Zone 1	Major	1	0	--	--
	Zone 1	Significant	49	0	--	--
	Zone 1	Minor	78	0	--	--
	Zone 2	Major	6	0	--	--
	Zone 2	Significant	159	4	2.5	1 in 40
	Zone 2	Minor	220	7	3.2	1 in 31
	Unclassified	Major	0	0	--	--
	Unclassified	Significant	4	0	--	--
	Unclassified	Minor	9	0	--	--
Oil Total			**526**	**11**	**2.1**	**1 in 48**
Gas	Zone 1	Major	22	0	--	--
	Zone 1	Significant	227	3	1.3	1 in 75
	Zone 1	Minor	106	6	5.7	1 in 18
	Zone 2	Major	88	0	--	--
	Zone 2	Significant	689	14	2.0	1 in 49
	Zone 2	Minor	353	21	5.9	1 in 17
	Unclassified	Major	4	0	--	--
	Unclassified	Significant	16	1	6.3	1 in 16
	Unclassified	Minor	21	1	4.8	1 in 21
Gas Total			**1526**	**46**	**3.0**	**1 in 33**
Condensate	Zone 1	Major	0	0	--	--
	Zone 1	Significant	10	0	--	--
	Zone 1	Minor	38	2	5.3	1 in 19
	Zone 2	Major	1	0	--	--
	Zone 2	Significant	46	0	--	--
	Zone 2	Minor	110	8	7.3	1 in 14
	Unclassified	Major	0	0	--	--
	Unclassified	Significant	1	1	100	1
	Unclassified	Minor	0	0	--	--
Condensate Total			**206**	**11**	**5.3**	**1 in 19**

Table 3.13 (continued)

Fluid Type	Area Classification	Release Size	Number of Releases	Number Ignited	% Ignited	Approximate Probability
2-phase	Zone 1	Major	8	0	--	--
	Zone 1	Significant	41	0	--	--
	Zone 1	Minor	11	0	--	--
	Zone 2	Major	21	0	--	--
	Zone 2	Significant	112	0	--	--
	Zone 2	Minor	27	0	--	--
	Unclassified	Major	2	0	--	--
	Unclassified	Significant	4	0	--	--
	Unclassified	Minor	0	0	--	--
2-phase Total			**226**	**0**	**--**	**--**
Non-process	Zone 1	Major	1	0	--	--
	Zone 1	Significant	14	1	7.1	1 in 14
	Zone 1	Minor	21	8	38.1	1 in 3
	Zone 2	Major	6	0	--	--
	Zone 2	Significant	86	16	18.6	1 in 5
	Zone 2	Minor	131	53	40.5	1 in 2
	Unclassified	Major	0	0	--	--
	Unclassified	Significant	18	2	11.1	1 in 9
	Unclassified	Minor	53	16	30.2	1 in 3
Non-process Total			**330**	**96**	**29.1**	**1 in 3**

Zone 1—area in which an explosive atmosphere is likely to occur occasionally in normal operation.

Zone 2—area in which an explosive atmosphere is not likely to occur occasionally in normal operation and if it does occur is likely to do so only infrequently and will exist for a short period only.

Unclassified—area not known to contain any concentrations of flammable vapor, gas, liquid, or dust in the atmosphere.

Other tables are included in Thyer (2005), such as one relating ignition of various utility fluids (diesel, methanol, etc.). However, the number of data points is not as significant, and the type of areas surrounding the point of release is not defined, so that data is considered to be less useful.

3.2.5 HSE/Gummer and Hawksworth—Hydrogen

In 2008, the HSE published a review on the subject of hydrogen ignition. The potential mechanisms for spontaneous ignition of hydrogen are discussed at some length but are not expressed in probabilistic terms. The potential mechanisms, and the opinion about each, were summarized in Chapter 1.

Of note is a review of 81 previous hydrogen releases, only 4 of which involved a delay in ignition between the start of the release and the time of ignition. It is presumed, but not known, whether the other incidents were also ignitions. If they were, then the probability of immediate ignition of hydrogen might be assumed to be very high, and delayed ignition of hydrogen might be presumed to be very low. This may be contrasted with the work by Dryer et al. in which the opposite might be concluded for immediate ignition probability.

3.2.6   Cawley/U.S. Bureau of Mines

This article (Cawley, 1988) reviews the potential for low-energy sources to ignite mixtures, specifically a mixture of 8.3% methane in air. Uniquely among the literature sources found, this document provides experimental probability of ignitions that in some cases are extremely low (under 10^{-6}). Results are provided for various types of circuits, as shown in Figure 3.2 and Figure 3.3.

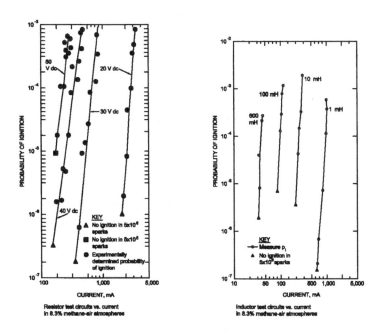

Figure 3.2. Probability of spark ignition (Cawley, 1988).

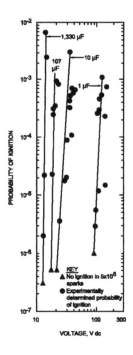

Figure 3.3. Probability of spark ignition for capacitor test circuits versus voltage in 8.3% methane-air atmospheres (Cawley, 1988).

3.2.7   Canvey

This document (HMSO, 1981) proposed the following probabilities of ignition for use in a risk assessment:

Release Onsite

Number of Ignition Sources	Ignition Probability
None	0.1
Very few	0.2
Few	0.5
Many	0.9

Release at Jetties

Probability of	..after a fire/explosion	...after a collision
Immediate ignition (within 30 seconds)	0.6	0.33
Delayed ignition (1/2 to a few minutes)	0.3	0.33
No ignition	0.1	0.33

Ignition of Cloud in Transit

Cloud Passes Over	Ignition Probability
Open land	0
Industrial site	0.9
Gas terminal	0.5

However, this information is dated and at the time was considered to be based on "judgment." Therefore, while this expert opinion might be considered, it will not be weighted greatly compared to some other sources.

3.2.8   Witcofski (NASA) Liquid Hydrogen

NASA (Witcofski, 1981) conducted seven releases of 5.7 m^3 liquid hydrogen in various ambient weather conditions for release durations ranging from 24 to 240 seconds. The purpose of the study was to evaluate the dispersion characteristics of the release, but the author noted too that "No attempt was made to ignite the spills and no spills ignited."

It is notable that the spills were directed through a long pipe into an earthen spill pond with no obvious ignition sources present either near the source or as far as the clouds dispersed (up to about 100 meters). This setup would remove some of the hydrogen ignition mechanisms that have been suggested in Chapter 1.

3.3   INFORMATION DEVELOPED BY INDUSTRY GROUPS

Information in one form or another suggested by industry groups follows next.

3.3.1   Cox/Lees/Ang

Perhaps the earliest work to quantify a relationship between release rate and ignition probability is an often-cited analysis from Cox, Lees, and Ang (1990). The proposed relationships are based on offshore event data and are somewhat dated, so they may not be reflective of current onshore operations. However, they are still widely used (Table 3.14).

Cox et al. (1990) also plots these points in a manner such that the following equations can be used:

$$POI_{gas} = 0.0156 \times Flow^{0.642}$$

(3-5)

$$POI_{liq} = 0.0131 \times Flow^{0.393}$$

(3-6)

where the flow is in kilograms per second.

This source also cites earlier work regarding the probability of explosion and proposes the following for a "standard plant" (Table 3.15).

Table 3.14. Estimated probability of ignition (Cox et al., 1990)

	Probability of Ignition	
Leak Size	Gas	Liquid
Minor (<1 kg/s)	0.01	0.01
Major (1–50 kg/s)	0.07	0.03
Massive (>50 kg/s)	0.3	0.08

Table 3.15. Estimated probability of explosion, given ignition (Cox et al., 1990)

Leak Size	Probability of Explosion, Given Ignition
Minor (<1 kg/s)	0.025
Massive (>50 kg/s)	0.25

Table 3.16 is a synopsis by Daycock and Rew (2004) of information from Cox et al. that was also developed from a review of incidents, although some information may be based on expert opinion rather than on actual data.

The 10^{-4} value in Table 3.16 illustrates the effect of very low release rates on ignition probabilities. However, it is below what most practitioners today would normally evaluate as a leak or release of concern or assume as a lower limit for POI and probably represents either very small events or other situations which should not be applied generally for process plant releases, even for small ones.

Table 3.16. Ignition probability estimates from various incident data reviews (Daycock and Rew 2004, derived from Cox et al., 1990)

Source	Type of release	Size of release	Location	Probability of ignition	Comments
Kletz (1977)	Polyethylene VCE	mostly small	general - on or near site	10^{-4}	good jet mixing with air
	Hydrogen & hydrocarbons mix	general	general - on or near site	0.033	
	(hot, @250bar)	> 10 ton		0.1 to 0.5	
Browning (1969)	LPG release	"massive"	general - on or near site	10^{-1}	Assuming no obvious source of ignition and
	Flammable liquid, flash-point <110°F	general		10^{-2}	explosion-proof electrical equipment.
	Flammable liquid, flashpoint 110-200 °F	general		10^{-3}	Multiply by 10 if strong ignition source present.
1st Canvey Report (1978)	LNG vapour clouds	"limited"	general - on or near site	10^{-1}	
		"large"		1	
2nd Canvey Report (1981)	LNG vapour clouds	general	on-site	0.1	"no" sources of ignition
				0.2	"v. few" sources of ignition
				0.5	"few" sources of ignition
				0.9	"many" sources of ignition
Dahl (1983)	Gas	blowouts	off-shore	0.3	based on 123 incidents
	Oil	(massive)		0.08	based on 12 incidents

3.3.2   E&P Forum

The E&P Forum (1996) collected much of the same offshore data that is reported elsewhere in this chapter. The POI tables are not repeated here, but Table 3.17 is from the World Offshore Accident Database (WOAD, 1994), which describes sources of ignition.

This is not to say that the distribution of offshore ignition sources will necessarily correspond to onshore ignition sources, but this may provide some insight on the relative strength of various types of ignition sources.

3.3.3   API RBI

The American Petroleum Institute's Risk-Based Inspection Standard (API, 2000) suggests the following ignition probabilities for refinery releases (Table 3.18 through Table 3.21). The benefits of these tables are that, although not explicitly derived from data, the information is derived for onshore facilities. Also, unlike almost all other references, the API tables account in detail for chemical type, state of the material releases, and type of ignition outcome.

Table 3.17. Distribution of ignition sources on platforms
(E&P Forum 1996, from WOAD 1994)

Ignition Type	Percentage
Electrical equipment	9%
Hot work	39%
Rotating machinery	26%
Exhaust	17%
Ignition by rupture	9%

Table 3.18. Specific event probabilities—continuous release
autoignition likely[a] (API, 2000)

Fluid	Final State Liquid—Processed Above AIT Probabilities of Outcomes					
	Ignition	VCE	Fireball	Flash Fire	Jet Fire	Pool Fire
C_1—C_2						
C_3—C_4						
C_5						
C_6—C_8	1				1	
C_9—C_{12}	1				1	
C_{13}—C_{16}	1				0.5	0.5
C_{17}—C_{25}	1				0.5	0.5
C_{25+}						1
H_2						
H_2S						
Fluid	Final State Gas—Processed Above AIT Probabilities of Outcomes					
	Ignition	VCE	Fireball	Flash Fire	Jet Fire	Pool Fire
C_1—C_2	0.7				0.7	
C_3—C_4	0.7				0.7	
C_5	0.7				0.7	
C_6—C_8	0.7				0.7	
C_9—C_{12}	0.7				0.7	
C_{13}—C_{16}						
C_{17}—C_{25}						
C_{25+}						
H_2	0.9				0.9	
H_2S	0.9				0.9	

[a] Must be processed at least 80°F above AIT.

Table 3.19. Specific event probabilities—instantaneous release autoignition likely[a] (API, 2000)

Fluid	Final State Liquid—Processed Above AIT Probabilities of Outcomes					
	Ignition	VCE	Fireball	Flash Fire	Jet Fire	Pool Fire
C_1—C_2	0.7		0.7			
C_3—C_4	0.7		0.7			
C_5	0.7		0.7			
C_6—C_8	0.7		0.7			
C_9—C_{12}	0.7		0.7			
C_{13}—C_{16}						
C_{17}—C_{25}						
C_{25+}						
H_2	0.9		0.9			
H_2S	0.9		0.9			
Fluid	Final State Gas—Processed Above AIT Probabilities of Outcomes					
	Ignition	VCE	Fireball	Flash Fire	Jet Fire	Pool Fire
C_1—C_2	0.7		0.7			
C_3—C_4	0.7		0.7			
C_5	0.7		0.7			
C_6—C_8	0.7		0.7			
C_9—C_{12}	0.7		0.7			
C_{13}—C_{16}						
C_{17}—C_{25}						
C_{25+}						
H_2	0.9		0.9			
H_2S	0.9		0.9			

[a] Must be processed at least 80°F above AIT.

Table 3.20. Specific event probabilities—continuous release
autoignition not likely[a] (API, 2000)

Fluid	Final State Liquid—Processed Below AIT Probabilities of Outcomes					
	Ignition	VCE	Fireball	Flash Fire	Jet Fire	Pool Fire
C_1—C_2						
C_3—C_4	0.1					
C_5	0.1				0.02	0.08
C_6—C_8	0.1				0.02	0.08
C_9—C_{12}	0.05				0.01	0.04
C_{13}—C_{16}	0.05				0.01	0.04
C_{17}—C_{25}	0.02				0.005	0.015
$C_{25}+$	0.02				0.005	0.015
H_2						
H_2S						
Fluid	Final State Gas—Processed Below AIT Probabilities of Outcome					
	Ignition	VCE	Fireball	Flash Fire	Jet Fire	Pool Fire
C_1—C_2	0.2	0.04		0.06	0.1	
C_3—C_4	0.1	0.03		0.02	0.05	
C_5	0.1	0.03		0.02	0.05	
C_6—C_8	0.1	0.03		0.02	0.05	
C_9—C_{12}	0.05	0.01		0.02	0.02	
C_{13}—C_{16}						
C_{17}—C_{25}						
$C_{25}+$	0.02					
H_2	0.9	0.4		0.4	0.1	
H_2S	0.9	0.4		0.4	0.2	

[a] Not likely if process temperature is less than autoignition temperature plus 80°F.

Table 3.21. Specific event probabilities—instantaneous release autoignition not likely[a] (API, 2000)

Fluid	Final State Liquid—Processed Below AIT Probabilities of Outcomes					
	Ignition	VCE	Fireball	Flash Fire	Jet Fire	Pool Fire
C_1—C_2						
C_3—C_4						
C_5	0.1					0.1
C_6—C_8	0.1					0.1
C_9—C_{12}	0.05					0.05
C_{13}—C_{16}	0.05					0.05
C_{17}—C_{25}	0.02					0.02
C_{25+}	0.02					0.02
H_2						
H_2S						
Fluid	Final State Gas—Processed Below AIT Probabilities of Outcome					
	Ignition	VCE	Fireball	Flash Fire	Jet Fire	Pool Fire
C_1—C_2	0.2	0.04	0.01	0.15		
C_3—C_4	0.1	0.02	0.01	0.07		
C_5	0.1	0.02	0.01	0.07		
C_6—C_8	0.1	0.02	0.01	0.07		
C_9—C_{12}	0.04	0.01	0.005	0.025		
C_{13}—C_{16}						
C_{17}—C_{25}						
C_{25+}						
H_2	0.9	0.4	0.1	0.4		
H_2S	0.9	0.4	0.1	0.4		

[a] Not likely if process temperature is less than autoignition temperature plus 80°F.

This data is useful in that it addresses the effect of material type in more detail than most sources and also discusses the effect of AIT. The 2009 edition of this publication provides expanded versions of the tables above.

3.3.4 API RP 2216

This publication (API, 2003) deals explicitly with the issue of ignition by hot surfaces in the open air, as opposed to the experimental values of AIT. As mentioned elsewhere, the API document notes the following:

- The surface temperature required to ignite a flammable release is typically much higher than the reported AIT.
- The probability that the release will ignite is related not only to the temperature of the hot surface but also to the contact area of the hot surface.
- The chances of ignition are greater the longer the contact is sustained.

Some interesting data is provided to support these statements (Table 3.22 and Table 3.23).

Table 3.22. Open-air autoignition tests under normal wind and convection current conditions (API, 2003)

Hydrocarbon	Published Ignition Temperatures (Approximate at Time of Test)		Hot-Surface Temperature without Ignition Occurring	
	°C	°F	°C	°F
Gasoline	280–425	540–800	540–725	1,000–1,335
Turbine Oil	370	700	650	1,200
Light Naphtha	330	625	650	1,200
Ethyl Ether	160	320	565	1,050

Table 3.23. Effect of ignition lag time on autoignition temperature (API, 2003)

Ignition Lag (s)	100		10		1	
Ignition Temperature	°C	°F	°C	°F	°C	°F
Hydrocarbon						
Pentane	215	419	297	567	413	775
Hexane	216	421	288	550	384	723
Heptane	202	396	259	498	332	630

Table 3.24 is also from this report and notes the effect of wind speed on open-air ignition temperatures:

**Table 3.24. Effect of wind velocity in autoignition tests
using kerosene (API, 2003)**

Wind Velocity Over the Hot Surface		Surface Temperature Required for Ignition	
m/s	ft/s	°C	°F
0.3	1.0	405	760
1.5	5.0	660	1,220
3.0	10.0	775	1,425

The publication goes on to mention that in offshore events hot-surface ignition was responsible for 35% of fires, "usually the exhaust system piping of an engine or turbine driver."

However, heretofore unmentioned hot surfaces such as high-pressure steam piping have been known to cause ignition of even heavier oils. Note that the temperature of saturated 600-psig steam is about 490 °F. Therefore, contact of flammables with the steam piping may be adequate to ignite many materials if the surface temperature exceeds the threshold for prolonged contact. Sustained contact would be expected with major releases.

3.3.5   IEEE

A group associated with IEEE evaluated the premise of the API tests in Section 3.3.4 as applied to an induction motor. The resulting paper (Hamer et al., 1999) describes the results of some tests on induction motors (3 and 20 hp) in which thermocouples were attached to the rotors to measure the temperature under both running and locked rotor conditions. The following conclusions were reached from these experiments:

- Ignition of a stationary hot rotor occurred at temperatures of 20–122 °C above the published AIT for the chemicals tested (diethyl ether, tetrafluoroethylene, hexane).

- Ignition of a running motor only occurred for diethyl ether at a temperature 69 °C above AIT. Note that the experiment was stopped at a temperature of about 350 °C, since a running rotor is considered to have a design temperature limit of 300 °C.

- Based on the above, Hamer et al. (p. 110) state, "there does not appear to be a significant risk of flammable vapor ignition by hot surfaces within an induction motor, except for the few materials with AITs below 200 °C."

In general terms, these results agree with API 2216 in that the observed hot-surface ignition temperature is higher than the AIT, and the ignition temperature is much higher under moving conditions (analogous to the high-wind-speed condition in the API tests).

It is also interesting to note that there may be circumstances, particularly for very low AIT materials, in which it would be undesirable to turn off a motor when a flammable release is detected.

3.3.6   UK Energy Institute

In 2006, the UK Energy Institute published an ignition probability model (UKEI, 2006) for use in QRAs with the co-sponsorship of the United Kingdom Offshore Operators Association (UKOOA) and the Health and Safety Executive. The model takes into account a large number of relevant variables and incorporates a dispersion model overlaid onto an input plant plot.   In the latter respects, its mandate is more ambitious than the scope of this book.

The other main differences between the UKEI effort and this book are the following:

- The UKEI model is focused mostly on offshore and LPG applications, although onshore process plants are also included.
- There is less energy spent on incorporating the fundamental physical-chemical properties of the materials being assessed than in this work. For example, users are told to assume a POI of 1 for releases above the AIT when, for reasons discussed in Chapter 1, this is not necessarily true (and could result in an underestimation of explosion risk). Simple corrections are also suggested for low-MIE materials, rather than the more elaborate correlations used here.

Bearing in mind the different focus of the two efforts, it remains to extract the approaches from the UKEI work that provide added insight or validation to the current effort. The notable findings from the UKEI project are discussed next.

The results are frequently compared to the Cox/Lees/Ang methods and, in general, the Cox/Lees/Ang methods result in much higher (factor of 1–10) ignition probabilities than the UKEI models. The authors of the UKEI book suggest that part of the reason may be more stringent controls being in place since the time of the Cox/Lees/Ang work.

The UKEI states that "... little or no new data have come to light in recent years ... the data do not allow specific aspects of ignition modeling such as the effect of ventilation, release type or location to be investigated."

Table 3.25 is proposed for the density of offsite ignition sources for onshore studies. In the larger context of ignition probability predictions, the differences by time of day do not seem great enough to incorporate in the model by requiring an additional input, except perhaps at Level 3.

Table 3.25. Ignition source densities for onshore offsite areas (UKEI, 2006)

	Industrial (per hectare)	Urban (per hectare)	Rural (per hectare)
Day	0.25	0.20	9.9×10^{-3}
Night	0.17	0.13	6.5×10^{-3}
Average	0.21	0.165	8.2×10^{-3}

Ignition source densities given per hectare. 1 hectare = 10,000 m^2.

The earlier HSE reports (Spencer, Daycock and Rew references) that are the basis of Table 3.25 provide day/night breakdowns for specific equipment types; these reports are further utilized in Table 3.26.

The effect of time on ignition is provided in Table 3.27, said to be based on onshore data. Note that the numbers are expressed as *relative cumulative* POIs and not in absolute terms. That is, the POI values are for periods of exposure up to and including the stated time and relative to the ignition probability at ~infinite (>1,000 seconds) exposure. Also consider the footnote, which is significant for the intended scope of this book.

It is probably reasonable to assume that the values above are appropriate for onsite ignition sources in localities having similar code requirements, and for offsite ignition sources in locations having similar land use requirements as the UK. The results might be significantly different in locations with different land use controls.

Table 3.26. Typical ignition source densities for onshore plant (UKEI, 2006)

Land Use Type	Ignition Sources	P	T_a	T_i	A	λ	M
Process Areas	Heavy equipment levels	0.5			1	0.028	50
Process Areas	Medium equipment levels	0.25			1	0.035	50
Process Areas	Light equipment levels	0.1			1	0.056	50
Storage	Material handling	0.1	10	20	0.333	0.0333	10
Offices	Light equipment levels	0.05			1	0.056	20

Table 3.27. Ignition timings overview (UKEI, 2006)

Plant Type	Relative Cumulative Probability of Ignition within Time, t(s)					
	1	10	30	100	1,000	> 1,000
Plant	0.22	0.29	0.36	0.63	0.94	1.0
Transport	0.53	0.53	0.53	0.60	0.86	1.0
Pipelines	0.24	0.30	0.31	0.39	0.61	1.0
CMPT/Blowouts	0.10			0.40	0.67	1.0
OIR12 Offshore[a]	0.89		0.92	0.97	0.99	1.0

[a]The OIR 12 data are dominated by small leak and ignition events that ignited "immediately"— most of these would not be relevant to MAH QRA. The "Plant" distribution in the table may be more representative of an ignition timing distribution for major leaks on an offshore installation.

With respect to release rate, the POI curves that are developed typically follow an "S" curve shape that may be a reasonable basis for a simplified yet accurate treatment of that variable. The plot in Figure 3.4 is an average of several plots from the UKEI report.

The curve in Figure 3.4 corresponds to Eq.(3-7) below with a lower bound of 0.001 and an upper bound of 0.3:

$$\text{Ignition Probability} = 0.4\,[1 - \exp(-0.005 \times \text{Flow Rate})] \qquad \textbf{(3-7)}$$

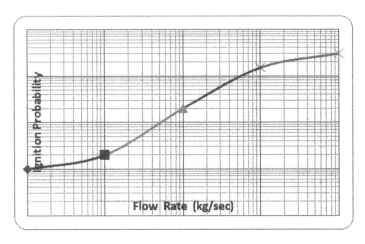

Figure 3.4. Typical results for large onshore plant—open-plant area (UKEI, 2006).

The report contains many such plots for different situations, and so this plot alone, while representative of the expected nature of the probability/rate relationship, should generally not be used as is.

3.4   INFORMATION DEVELOPED IN ACADEMIA

3.4.1   Ronza et al.

Ronza et al. (2007) first considered the existing literature and summarized the variables that other authors had determined affected ignition and explosion probabilities. These variables were:

- Amount spilled/released
- Material properties
- Type of accident
- Density of ignition sources
- Prevailing meteorology (wind speed or atmospheric stability)

The first, second, and last of these could be presumed to determine the size of a flammable cloud. The third and fourth items might determine the presence and type of ignition sources available, and the second item might determine the energy required for the ignition to occur.

Ronza et al. reached a conclusion similar to one of the driving forces for this book: "It is not always obvious whether figures are the result of expert judgment or historical analysis…in most cases it must be induced that the authors used these [historical data] in combination with expert judgment."

The specific interest of Ronza et al. was related to transportation applications. To this end, they utilized two U.S. government databases:

- The Hazardous Materials Incident Reporting System (HMIRS) and
- The Marine Investigation Module (MINMOD), aka Marine Casualty and Pollution Database.

The analysis of this database resulted in some potentially useful relationships, some of which are shown in Figure 3.5 through Figure 3.8.

In Figure 3.5 through Figure 3.7, the Y axis is the total probability of ignition, calculated as P_1 (probability of immediate ignition) plus the probability of failure to immediately ignite ($\bar{P}_1 = 1 - P_1$) multiplied by P_2 (the probability of delayed ignition given that immediate ignition does not occur). Figure 3.8 shows P_3, which is the probability of explosion.

The figures reinforce the concept that release magnitude (in this case expressed as amount released rather than release rate) affects the ignition probability, as does the type of material being released. Note that while the plots of ignition probability as a function of released chemical in Figure 3.8 are based on the flash temperature, it is also possible that the ignition may also be related to one or more other surrogate chemical properties such as molecular weight, boiling point, minimum ignition energy, etc.

There is no assessment of the prevalence of ignition sources; however, this source provides valuable insights for "typical" transportation events and might also be considered in relative terms for effects of known variables on the probability of ignition/explosion for process plant events.

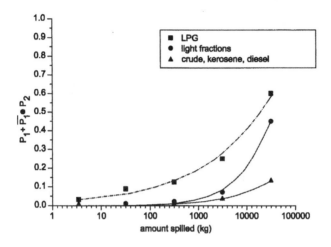

Figure 3.5. HMIRS total ignition probability data as a function of the hydrocarbon and the amount spilled (Ronza et al., 2007).

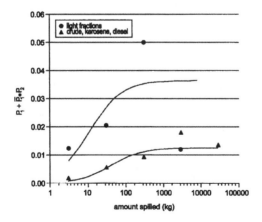

Figure 3.6. MINMOD total ignition probability data as a function of the hydrocarbon and the amount spilled (Ronza et al., 2007).

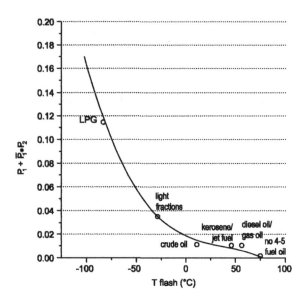

Figure 3.7. Average HMIRS total ignition probability data as a function of the
average flash point temperature of the hydrocarbon spilled
(Ronza et al., 2007).

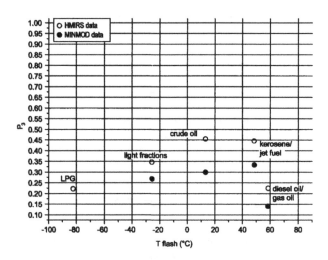

Figure 3.8. Average HMIRS and MINMOD explosion probability data as a
function of the average flash temperature of the hydrocarbon spilled
(Ronza et al., 2007).

3.4.2   Offshore Explosions (Loughborough)

A group at Loughborough University (Foster and Andrews, 1999) also investigated offshore ignitions. They considered a number of variables affecting explosion frequency, many of which will not be applicable to the typical onshore process plant facility:

- Frequency of leakage
- Frequency of ignition
- Time of detection of the gas concentration
- Availability of explosion-mitigating systems
- Reaction time of explosion-mitigating systems
- Availability of a blow-down system
- Availability of a gas detection system
- Availability of an isolation system
- Source strength of leakage
- Pressure-time history of the section
- Ventilation rate
- Concentration-time history of the gas-air mixture

These are incorporated into an event tree, and the results are calculated in principle through one or more lengthy equations composed of numerous integrals. For the purposes of this book, it is considered that the proposed methods are too complicated to be presented in a transparent way to a general audience of subject matter experts. In addition, the application and assumed surroundings are different from the primary process plant application of this book and do not incorporate some variables of interest to the current effort. Therefore, the work of Foster and Andrews is not directly utilized in developing the ignition probability algorithms in this book. However, it should be noted that many of the variables above are relevant to specific onshore applications, particularly indoor operations.

3.4.3   Srekl and Golob

Investigators from Slovenia (Srekl and Golob, 2009) reviewed local fire statistics to evaluate the frequency of fires in buildings involving a broad spectrum of activities, most of which were not industrial applications. Interestingly, they found no correlation between the amount of flammable materials present and the fire frequency. They ascribe this to strict regulations in the Republic of Slovenia.

They found that the predicted number of fires *per 10-year period* (*Nf*) can be predicted as follows:

$$Nf = 0.12 \times \text{Log\_hot} + 0.42 \times \text{Time} \qquad \text{(3-8)}$$

where:

Log_hot   =   the logarithm of the number and exposure time for the sum of ignition sources available

Time   =   the number of hours per day the building is active

3.4.4   Duarte et al.

The subject of ignition by hot surfaces was investigated at length by a team from Brazil (Duarte et al., 1998). The heat transfer that would result in an ignition from a hot surface was related to the flow of heated vapor away from a hot surface, which was used to explain why other observers have found that hot-surface temperatures need to be substantially higher than the report AIT in order for ignition to occur. These investigators showed that confining the flammable gas by a hot surface (Figure 3.9) inhibited this flow of heated gas and as a consequence led to lower ignition temperatures.

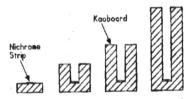

Figure 3.9. Experimental configurations used to test the effect of confinement on ignition temperature by a hot surface (Duarte et al.).

Table 3.28. Ignition temperature of propane for the configurations shown in Figure 3.9 (Duarte et al.)

Configurations	Ignition Temperature °C
Open	880–900
10 mm deep	870–890
20 mm deep	850–870
40 mm deep	800–820

3.4.5   Swain—Ignition of Hydrogen

Swain et al. (2007) experimentally investigated the ignition of releases of hydrogen. The results from this study suggested that the effects of increased pressure on increasing ignition probability (as relates to static discharge potential or increased overall flow rate) may not be as great as one might expect. The phenomenon of interest was whether an incipient ignition would be "blown off" the end of a flammable jet if the flame speed of the material was less than the jet velocity, and the test results seemed to confirm this:

- Releases at Mach 0.1 velocity ignited farther away from the source than releases of the same nominal flow rate (using a smaller diameter orifice) at Mach 0.2.

- The concentration of hydrogen at the further distance where ignition took place was approximately 7 vol.% at Mach 0.1 but 10 vol.% at Mach 0.2.

Since it is unlikely that there will be "hard" ignition sources in an operating plant at the distances used in the study (approx. 5 feet), this may infer that the dominant mode of hydrogen jet ignitions near the source may be static. It may also suggest that the probability of delayed ignition (and not just the probability of explosion) may be related to the degree of congestion/confinement in the surrounding plant—that is, that interruption of a jet by an obstruction allows the jet velocity to drop to a point where the flame speed can overcome it and reverse back toward the source.

3.4.6   Dryer et al.—Hydrogen and Light Hydrocarbons

Dryer et al. (2007) performed a series of experiments with hydrogen (and to a much lesser extent, natural gas) in which the presence of obstructions or confinement at the point of release (in this case, tubing/fittings downstream of a hydrogen cylinder rupture disk) can influence the odds of ignition. The presumed principles causing this combustion are described in Chapter 1.

The ignition phenomenon was only observed for release pressures greater than about 200 psig. The requisite confinement was effective down to a discharge piping length of about 1.5 inches, below which ignition did not occur. Ignition also disappeared at lengths greater than ~40 inches, which was attributed to combustion heat removal by the piping.

Also of interest was the fact that tests of releases into open atmospheres did not produce ignition at failure pressures as high as 800 psig. The conclusions might well be different at the elevated temperatures encountered in many hydrotreating applications. However, the implications for release of high-pressure streams with low ignition temperatures are significant—that ignition may occur in releases taking place in confined spaces (e.g., relief device discharge piping or perhaps flange leaks), but not into open spaces (e.g., pinhole leak in process vessel).

3.4.7   Britton—Silanes and Chlorosilanes

Britton (1990a) conducted a review of the hazards of silanes and chlorosilanes that provides some insights beyond those associated with this particular chemical class:

Release Pressure/Velocity—It was presupposed based on anecdotes and best judgment that increased discharge velocity was universally associated with increased ignition potential, due to both static electricity buildup and the simple fact that, for a given hole size, a higher discharge equals a larger cloud that could encounter more ignition sources. However, Britton reports work of his and another that showed that there are also mechanisms under which the opposite is true. At high velocities, the flame may blow off (that is, be extinguished when its discharge velocity exceeds the reversing flame speed). In addition, there were observations of silane mixtures in which ignition at the source did not occur until the release had been nearly stopped. This was attributed to the presence of pyrophoric compounds at the point of discharge that were only exposed to air when the release velocity was low enough to allow air migration back to the source.

Velocity/Temperature/Orifice Size—Figure 3.10 shows the impact of release conditions on silane ignition. These effects were only marginally repeatable, however, and the relationships are complex enough that they cannot be rigorously taken into account in a general ignition probability model. However, they do validate the observations of others that these variables can be important to ignition probabilities.

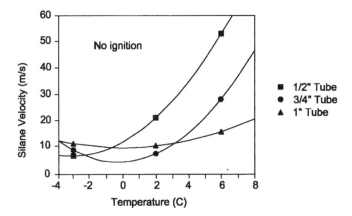

Figure 3.10. Critical ignition velocities of silane at different temperatures (Britton, 1990).

3.4.8   Pesce et al.

In this paper, the authors use the UK Energy Institute algorithms as a starting point and go on to refine and modify this and other earlier works (such as Rew et al.) to develop the immediate and delayed ignition rule sets given in Table 3.29.

Table 3.29. Minimum immediate ignition probability

Substance Group	$T_R \leq (T_{AI} - 27\ °C)$		
	P ≤ 2 barg	2 barg > P > 50 barg	P ≥ 50 barg
IIA	5%	10%	20%
IIB	10%	20%	30%
IIC	20%	30%	40%

Substance Group	$(T_{AI} - 27\ °C) < T_R < (T_{AI} + 27\ °C)$		
	P ≤ 2 barg	2 barg > P > 50 barg	P ≥ 50 barg
IIA	60%	70%	80%
IIB	70%	80%	90%
IIC	80%	90%	95%

The "substance groups" above are defined by an Italian classification code, "P" and "T_R" are the process pressure and temperature, respectively, and "T_{AI}" is the autoignition temperature. Note that the value of 27 °C in the table is a misconversion of the API autoignition/process temperature adjustment of 80 °F; the 27 °C value should read 44 °C instead.

The authors also discuss ignition source "efficiency" as a function of both the ignition source strength and the substance group for "standard" ignition sources and for hot surfaces (as interpreted from API 2216): See Tables 3.30 and 3.31.

Table 3.30. Efficiency factors for ignition sources that generate sparks

Substance Group	Ignition Source Strength				
	Certain	Strong	Medium	Weak	Negligible
IIA	1	0.60	0.05	0.01	0
IIB	1	0.75	0.27	0.025	0
IIC	1	0.90	0.50	0.04	0

Table 3.31. Efficiency factors for hot surfaces

Temperature	Efficiency Factor Value
$T_S < T_{AI}$	$\varepsilon = 0$
$T_{AI} \le T_S \le (T_{AI} + 105 \,°C)$	$\varepsilon = 0.5$
$T_S > (T_{AI} + 105 \,°C)$	$\varepsilon = 1$

In Table 3.31 "T_S" is the temperature of the hot surface.

3.5   INFORMATION DEVELOPED BY INDIVIDUAL COMPANIES

The information that follows was prepared by individual companies active in the ignition probability/risk assessment field.

3.5.1   Spouge

A publication (Spouge, 1999) related to *offshore* risk assessment provides suggested probabilities of ignition obtained from other sources (Table 3.32) but based on actual data.

Table 3.32. Expected ignition probabilities for offshore oil and gas leaks (Spouge, 1999)

Release Rate Category	Release Rate (kg/s)	Gas Leak POI	Oil Leak POI
Tiny	<0.5	0.005	0.03
Small	0.5–5	0.04	0.04
Medium	5–25	0.10	0.06
Large	25–200	0.30	0.08
Massive	>200	0.50	0.10

A "judgmental" suggestion was proposed (Table 3.33) to account for the effect of time, while noting that it agreed with blowout data but not HSE data.

Table 3.33. Offshore ignition delay probabilities (Spouge, 1999)

Time Interval (minutes)	POI in Interval	POI by End of Interval
0 (immediate)	0.10	0.10
0–5	0.20	0.30
5–20	0.37	0.67
20–60	0.29	0.96
>60	0.04	1.00

3.5.2 Moosemiller

Work performed for an industry consortium group (Moosemiller, 2010) incorporated some of the previous literature along with some proposed algorithms to cover variables not previously addressed in the literature but considered to be significant variables. The resulting relationships were proposed as follows:

3.5.2.1 Probability of Immediate Ignition

$$P_{imm.ign.} = [1 - 5000e^{-9.5(T/AIT)}] + [0.0024 \times (P)^{1/3} / (MIE)^{2/3}]$$

The autoignition temperature (AIT) and process temperature (T) are in degrees Fahrenheit, the process pressure (P) is in psig, and MIE is in millijoules. The following constraints are placed on this equation:

- Minimum value of 0 is allowed for T
- For $T/AIT < 0.9$, P_{ai} (ignition probability attributable to autoignition) = 0
- For $T/AIT > 1.2$, $P_{ai} = 1$
- $P_{imm.ign.}$ cannot be greater than 1

The first term in the equation above represents the contribution of autoignition (recognizing that reported autoignition temperatures are imprecise) and static discharge from some source at or near the point of release.

3.5.2.2 Probability of Delayed Ignition (PODI)

This probability is based on a default starting point of 0.3, which is then modified by the following factors:

Material Factor $M_{mat} = 0.6 - 0.85 \log(MIE)$, with an upper limit of 3.

Magnitude Factor $M_{mag} = 7 \times e^{[0.642 \times \ln (FR) - 4.67]}$, where FR is the flow rate in pounds/second (derived from Cox/Lees/Ang), with an upper limit of 2.

Duration Factor $M_{dur} = [1 - (1 - S^2) \times e^{-(0.015 \times S)t}] / 0.3$, when t is expressed in seconds, and the "strength" of the ignition source is as adopted from the works by Spencer, Daycock, and Rew.

Indoors Factor $M_{in/out} = 2$ if the operation is inside; 1 if outside

A more speculative indoor model seeks to take into account building variables that are considered to be relevant, such as the air change rate, ventilation orientation, and area electrical classification. This is considered in the present analysis in Appendix B.

The default and multipliers are handled mathematically as follows to ensure that the resulting PODI falls between 0 and 1:

If the product of the multipliers is >1, then $P_{del.ign.} = 1 - (0.7/\prod M_i)$

If the product of the multipliers is <1, then $P_{del.ign.} = 0.3 \times \prod M_i$

In the equations above, the "M_i" terms represent the individual multipliers, such as M_{mag}, M_{dur}, etc.

3.5.2.3 Probability of Explosion, Given Delayed Ignition

After Cox/Lees/Ang, the probability of explosion (given ignition) is based on the flow rate but is modified based on the explosive tendency of the chemicals being released, as follows:

"Low reactivity"—Multiply by 0.3

"Medium reactivity"—No modifier

"High reactivity"—Multiply by 3

The resulting upper limit to $P_{exp/g/ign}$ must be 1. This last multiplier is not based on "data"; rather, it is based on general explosion modeling principles where chemical "reactivity" (that is, fundamental burning velocity) drives a flame front toward explosivity.

3.5.3   Johnson—Humans as Electrostatic Ignition Sources

Johnson (1980) noted that several fairly restrictive features need to be present for ignition of flammable clouds by human electrostatic discharge to be realized. One factor is a gas concentration in a fairly narrow range. This is illustrated in Figure 3.11, although it is not clear if the narrow range criterion applies to all fuels or not. The form of this figure is inferred from a very limited number of data points.

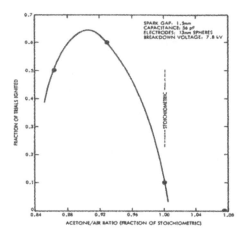

Figure 3.11. Electrostatic ignition probability as function of acetone/air ratio (Johnson, 1980).

A low absolute humidity is also required; Johnson suggests that a relative humidity of 60% at 70 °F is sufficient to prevent ignition. Otherwise, it was concluded that any chemicals or mixtures with an MIE of under 5 mJ could be ignited by this mechanism.

Figure 3.12 also illustrates the qualitative differences between "conventional" sparks and human spark ignition. The extreme dependence of ignition probability on ignition energy is similar to that observed in the work by Cawley (1988).

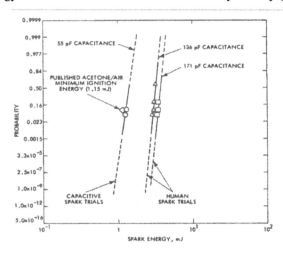

Figure 3.12. Probit curves for acetone/air spark ignition.

3.5.4   Jallais—Hydrogen

Air Liquide (Jallais, 2010) did a review of hydrogen ignition probability work that had been reported by others. Jallais noted the effect of release rates on ignition probability for chemicals in general and also noted the fairly wide range of predictions that have been suggested for hydrogen specifically.

Eliminating the high and low values from each category to remove the extremes, a geometric average of the remainder results in the following:

(for hydrogen at 20 °C, 200 bars, 1 kg/s release rate, unconfined for 1000 seconds into obstructed area)

P immediate = 0.25

P delayed = 0.15

3.5.5   Zalosh—Hydrogen

Factory Mutual (Zalosh et al., 1978) performed an analysis of reported hydrogen fire and explosion events for the U.S. government. Table 3.34 shows the breakdowns recorded.

Of course, there is significant potential for data bias in these results. One can argue that it is probably less likely that unignited events would have been reported and that it is more likely that explosions would have been reported than fires. Therefore, this data should be utilized with great care and not to the exclusion of other sources of hydrogen ignition data.

Table 3.34. Distribution and type of hydrogen incident

Type of Incident	Number of Incidents	Fraction of Total
Fire	74	0.264
Explosion	165	0.589
Pressure Rupture	12	0.043
Unignited Release	20	0.071
Fire and Explosion	3	0.011
Other	6	0.021
TOTAL	**280**	

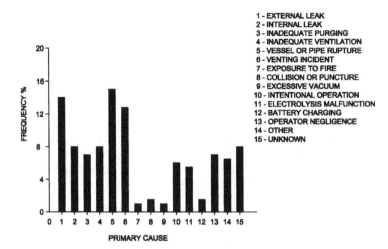

Figure 3.13. Hydrogen incident statistics (1961–1977) distribution by primary cause (Zalosh et al., 1978).

Zalosh et al. go on to provide a breakdown by cause, as shown in Figure 3.14.

These events include many that may not be of direct applicability to users of this book.

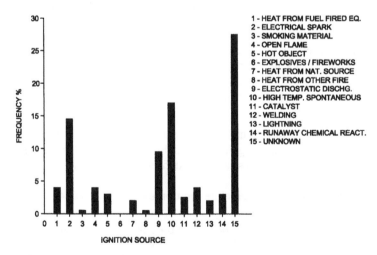

Figure 3.14. Hydrogen incident statistics (1961–1977) distribution by ignition source (Zalosh et al., 1978).

3.5.6   Smith—Pipelines

Smith (2011) did a comparison of ignition and explosion probabilities in petroleum versus natural gas pipelines. He considered the "environmental" aspects of the two pipeline applications to be similar, in that both are well-regulated and have similar construction and operating practices. They also have a relatively large body of data to draw upon.

The study attempted to remove any data that might represent a significant difference in operation between the two pipeline types. This resulted in deleting pipelines that were offshore, under water, or below ground. Events in buildings, tanks, or under structures were also eliminated. The release and ignition data remaining after the sorting process are shown in Table 3.35.

For the purposes of this book there are still complications in the use of this data. For example, while we know the total number of ignitions, the proportion of the ignitions that are "immediate" versus "delayed" cannot be determined. For the same reason, the fraction of delayed ignitions that resulted in explosions cannot be calculated. Nonetheless, the "final" ignition/explosion fractions are known, and they tell a clear story—that natural gas releases are much more likely to ignite than liquid petroleum releases.

After discounting some possible explanations for the disparity, Smith concludes that the greater number of ignitions in natural gas pipelines is due to differences in static electricity formation. One variable unmentioned in the Smith analysis that could have a bearing on these conclusions is that of ignition energy. The components of natural gas have an ignition energy of about 0.27 mJ. The composition of "petroleum liquids" pipeline contents is potentially quite varied, having ignition energies ranging from 0.25 mJ to tens or hundreds of millijoules, not even accounting for the energy required for dispersal of the liquid into a spray capable of being ignited. Thus there is a very real possibility that liquid contents of pipelines should be harder to ignite on an energy basis, irrespective of static issues.

These results are directionally consistent with almost all other sources of petroleum liquid vs. gas ignition probability comparisons, even if the mechanism(s) for the differences cannot be definitely determined.

Table 3.35. "Sorted" U.S. pipeline ignition events (Mar. 2004–Nov. 2010)

	Number of Incidents	Number of Ignitions	Number of Explosions
Petroleum Liquids	474	24	9
Natural Gas	386	293	29

3.6   STUDIES SPECIFIC TO IGNITION OF SPRAYS

Not surprisingly, most of the literature related to ignition of flammable liquid sprays is related to the study of internal combustion engines. Examples include Liao (1992), Wehe and Ashgriz (1992), and Lee et al. (1996). In most respects, such work (of ignitions within small combustion chambers) is not relevant to the application of interest in this book (open field ignitions). Nonetheless, there are findings of these ignition studies that may be applied to this work, which are discussed next.

3.6.1   Lee et al.

Lee et al. (1996) investigated the ignition of multicomponent fuel sprays—in this case, mixtures of *n*-heptane and *n*-decane. The findings that may be significant for ignitions in an external environment include the following:

Minimum Ignition Energies Increase with Increasing Droplet Diameter—This result may seem intuitive but is still worth considering (Figure 3.15).

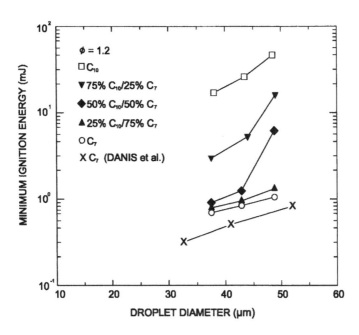

Figure 3.15. Minimum ignition energies versus droplet diameter for bicomponent fuel sprays (Lee et al.).

Since there is no expectation of a user to provide droplet sizes as an input to the ignition probability algorithm, the relevance of this finding to the effort in this book is with respect to other parameters that affect droplet size. For chemicals that do not vaporize significantly upon release, the most significant parameter is source pressure. Pressure is already incorporated in the models because of its influence on release rate and static discharge; this is another factor to consider.

Small Amount of Low-MIE Impurities Can Disproportionately Lower the MIE of a Mixture—In this respect, MIE is like the physical properties of mixtures whose behavior follows LeChatalier mixing rules:

$$MIE_{mix} = 1/(\sum (x_i/MIE_i))$$

where x_i = mole fraction of component "i" and MIE_i = MIE of component "i."

This is illustrated in Figure 3.16.

The stars in the figure are superimposed, and indicate what use of the LeChatalier mixing rule would predict for the 44-µm cases. The result is not a perfect match but shows that LeChatalier's rule may be close enough for the users of this book.

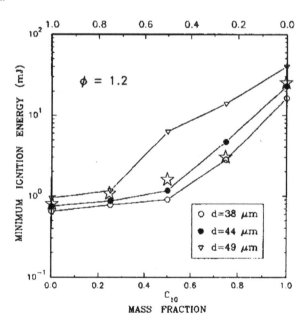

Figure 3.16. Minimum ignition energies versus component mass fraction for bicomponent fuel sprays (Lee et al.).

3.6.2   Babrauskas

Babrauskas (2003) reported on the ignition temperature of sprays as a function of droplet size and of temperature (Figure 3.17 and Figure 3.18).

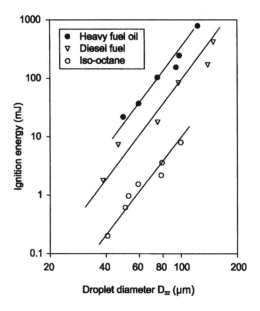

Figure 3.17. Effect of droplet diameter on MIE of aerosols with larger droplet sizes (Babrauskas, 2003).

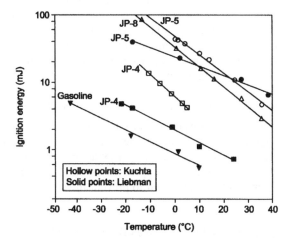

Figure 3.18. The MIE for various sprays (Babrauskas, 2003).

Figure 3.17 confirms the trend shown in Lee et al. Figure 3.18 has implications for whether the MIE should be adjusted for process temperature.

3.7   CASE HISTORIES

Most of the ignition case histories in the literature that are discussed in detail involve an event in which a "standard" operation (e.g., filling a tank) is being conducted such that the fuel is always present, but with some deviation from the intended operation (e.g., an accumulation of static charge) that results in an ignition.

In contrast, the focus of this book is on situations in which the ignition sources may always be present and where some event results in the fuel being exposed to the ignition sources. Thus, most major release event histories note whether the release ignited or not but do not look at the issue further in any sort of probabilistic way.

The following case histories provide examples from this relatively limited literature.

3.7.1   Britton—External Ignition Events

Britton (1999) relates the following cases that do not provide quantification of ignition probabilities but do provide insights into a variety of ignition situations that might otherwise be overlooked in a review of the standard POI literature.

3.7.1.1   Clean Rooms

Britton notes that "clean rooms" are often maintained as low-humidity environments which can promote static accumulation and thus increase the chances of ignition if a release takes place. The safeguards in this case are the use of antistatic floors as well as grounding of any metal equipment/storage and avoidance of plastic equipment. Personnel should be grounded as well, and clothing should be of an antistatic type.

3.7.1.2   Gas Releases "Contaminated" with Another Phase

The presence of pipe scale or suspended liquids in a gas stream can create charged particles that can ignite a release. Such a phenomenon can even occur as the result of initiation of a water spray or carbon dioxide intended to *prevent* a fire. However, in the former case, fire suppression systems can be provided with appropriate grounding and nozzle design to prevent them from becoming a source of ignition.

3.7.1.3   Self-Ignition of Released Gases Below the AIT

Materials such as hydrogen that have low ignition energies may be ignited when near to the point of being discharged to the atmosphere. Various models have been proposed to account for such ignitions in typical process release conditions, including:

- Corona-brush discharges.
- Hydrogen releases:
  - From accumulation of dihydrogen cations on pipe walls due to reaction of hydrogen with the pipe
  - From release of a gas stream that has entrained particulates (such as reduced metal oxides) that are pyrophoric.

Britton notes in passing that an ignition mechanism sometimes incorrectly ascribed to hydrogen ignitions is the inverse Joule-Thompson effect. In fact, the temperature rise resulting from this effect in hydrogen is modest, so that it would only matter if the hydrogen was already very near its AIT.

3.7.1.4   Ball Valve Stem Leaks

Britton also comments on the potential for ignition of packing leaks in ball valves, since (a) some ball valve designs can isolate the valve stem from the valve body and (b) ball valves are frequently used in services such as acetylene, a chemical which is relatively easily ignited. Acetylene poses a secondary hazard as well—if a fire begins a detonation inside the piping, this could result in autodecomposition of the acetylene.

The above condition can be taken into account in the ignition probability algorithms to the extent that the low ignition temperature of acetylene is applied. It is expected that this particular ball valve mechanism is a small subset of acetylene releases, in general. And so, while this possibility is an instructive lesson to the reader, it is assumed that the small benefit of adding a ball valve/acetylene parameter to account for (small) stem leaks in the POI prediction algorithms is offset by the effort needed to add one or two additional data entries.

3.7.1.5   Gas Leaks Under Insulation

The last case relevant to external ignition events related by Britton is the case of a pressurized gas release under insulation that loosens the insulation and pushes it against metal weather barriers, which as a result may develop a charge. As with the previous example it is useful to know of this issue, but for the purposes of developing algorithms it is a complexity with relatively small benefit in predicting POI. It would also be problematic to develop a defensible basis for quantifying its effect.

3.7.2   Pratt—Gas Well and Pipeline Blowouts

Pratt (2000, p.151) relates the postulated basis for observed ignitions in well and pipeline blowouts. This can be instructive, since much of the seminal POI data was collected for gas/oil production blowouts, and it would be useful to evaluate the differences in that application and onshore production releases with respect to the factors that lead to ignition.

The proposed mechanism for ignition is the high-velocity discharge of a mixed gas/liquid stream from a pipe that allows a highly charged gas/droplet cloud to form—it is presumed that "the competing forces of mutual electrostatic repulsion between the aerosol droplets and the mobility of the aerosol droplets through the air are such that the expansion of the cloud is sufficiently retarded to allow the buildup of an electric field." Since air and metallic grounded equipment are also present, a brush discharge and ignition can result. Blowouts also are likely to eject various kinds of solid particles that can serve to promote ignition.

In contrast to the two-phase release case discussed to this point, single-phase gas leaving a pipe will not have a charge except, perhaps, if it is accompanied by scale or other particulates.

3.7.3   Gummer and Hawksworth—Hydrogen Events

The following is quoted from Gummer and Hawksworth (2008, pp. 2-4).

The 1922 Incident Investigation

This incident and the subsequent investigation and research work was reported in Engineering, from work undertaken by Nusselt in Germany. After several spontaneous ignitions of hydrogen at 2.1 MPa being discharged to atmosphere had been reported, work was undertaken to determine the cause. Various experiments were undertaken on discharging hydrogen to atmosphere, but no ignitions occurred despite discharging through many different types of nozzle made from differing materials. However, cylinders had been noted for having quantities of iron oxide (rust) in them even though they were apparently dry, and it was thought that there was potential for electrostatic charging to occur. Despite many differing finely powdered materials being used, no ignitions occurred except for extremely finely ground iron oxide. Manganese dioxide also caused ignition, so it was thought that the rust was catalysing the oxidation of the hydrogen. Therefore, mixtures of hydrogen and oxygen were stored at an initial pressure of 1.1 MPa at various temperatures in the presence of iron oxide to determine whether the oxide catalysed the reaction. At ambient temperatures, no pressure changes occurred, even after a few weeks, but at temperatures above ambient, the pressure slowly fell, indicating that the oxidation reaction was occurring. The times were about 24 hours at 100 °C, 9 hours at 200 °C, and one hour at 380 °C. There was no explosion at any time.

Subsequent experiments on discharging hydrogen into an open funnel fitted with a long pipe showed no ignitions except when the funnel was obstructed by an iron cap. The mechanism was not understood, so further trials were undertaken. Only when the trials were undertaken in the dark was a corona discharge observed. This was revealed when the hydrogen leaked out of a flange—the corona discharge was visible, which increased when the pipe was tapped to stir up dust. An ignition followed after the tapping. Further work showed that when sharpened copper wires were used to promote corona discharges, ignition occurred when the point was bent away from the gas direction, whereas no ignition occurred when the wire was pointing in the direction of flow. Consequently, it is apparent that a corona discharge was likely to have been the source of ignition in this case.

The 1926 and 1930 Incidents and Experiments

The first incident occurred in 1926, but was only reported by Fenning and Cotton in 1930 after a second explosion occurred. As the cause of the ignition in both cases was obscure, experimental work was undertaken to try to establish the mechanism. The second explosion occurred when the isolation valve between a pressurised pipeline and a chromium plated vessel was opened to depressurise the line from about 4.9 MPa. The explosion occurred immediately, and traces of water were found in the previously dry vessel confirming that combustion had taken place. It was noticed that there was ample evidence of fine dust, presumably metal oxide, being present in the pipe-work during the examination after the explosion. This led Fenning and Cotton to surmise that the explosion had been initiated by an electrostatic discharge, presumed to have been generated by the fine dust being blown along the pipe by the high velocity hydrogen. However, despite many attempts, no ignition was achieved in their experiments.

The first explosion was not investigated at the time, but as the same workers were involved, they reviewed the circumstances. In this first explosion, the pressure was only about 6.6 kPa above atmospheric pressure in a glass vessel. No obvious source of ignition was present, but it was observed that a fine jet or spray of mercury may have been projected into the gas mixture. The mixture was said to have been "...a sample of a 'complete combustion' hydrogen-air mixture...", which can be taken to mean a stoichiometric mixture. Again, there is a suggestion that an electrostatic ignition mechanism is possible.

Incidents Reported by Bond

Bond reports two incidents, sourced from a private conversation, where hydrogen ignited. In the first incident, hydrogen at a pressure of 11.1 MPa was leaking from a gasket between two flanges. The hydrogen had not ignited at the time when the fitters arrived to tighten the bolts. It was reported that on the second strike of the hammer wrench that was being used to tighten the bolts, there was an ignition. It is not apparent whether the ignition source was an impact spark from a hammer wrench being used to tighten the bolts on the joint, or attributed to the mechanism of a diffusion ignition. The second incident refers to a cylinder of hydrogen being connected to a piece of laboratory apparatus. The laboratory technician cracked the valve open to clear any dirt out of the connection, and when he did so, the escaping gas ignited immediately. Bond attributes this ignition to the phenomenon of diffusion ignition. Whilst no pressure of gas is quoted in this second incident, it can be assumed that the pressure would have been the typical full cylinder pressure of 23 MPa.

Jackass Flats Incident, 1964

This incident, reported by Reider, Otway and Knight, involved the deliberate release of a large quantity of hydrogen to determine the sound pressure levels. The hydrogen was released from storage at an initial pressure of 23.6 MPa and an initial rate of 54.4 kg s^{-1}, for a period of 10 seconds. The gas was transferred through a 200 mm nominal bore pipe and a 150 mm bore ball valve to a cylindrical vessel fitted with a convergent-divergent nozzle venting to atmosphere. The intention was to discharge the gas without combustion and again with deliberate combustion, so that the sound level due to the combustion could be measured. In the run where the gas was not deliberately ignited, after 10 seconds, the 150 mm diameter valve was being closed, and three seconds after starting to close the valve, ignition occurred.

Prior to the experimental discharge, three potential ignition mechanisms were examined, as it was recognised that ignition during a "non-ignition" would require the run to be aborted. The three potential ignition mechanisms examined were electrification of the gas; electrification of the particles in the gas; and metal particles abrading a metal bar welded across the mouth of the nozzle. Of these, the first was discounted as pure gases are known to have negligible electrostatic charging. The second was considered, but as the system had been thoroughly cleaned out and blown down prior to the test, it was considered that there would not be any particles present. However, the velocity of the gas being discharged, at 1216 m s^{-1}, was far higher during the run than had been used before, so this potential mechanism could not be discounted. The third mechanism was considered as a possibility as the discharge velocity was high—possibly dislodging particles, and impacting them on the bar. This too could not be discounted. After the ignition, it was found that the bar had been torn loose at one end, and this may have presented a possible ignition source which had not been foreseen.

4 ADDITIONAL EXAMPLES

4.1 INTRODUCTION TO EXAMPLES AND POTENTIAL "LESSONS LEARNED"

The following examples provide some straightforward situations expected to be typical of situations encountered by users. There are no special complications involved, and the information needed should be readily available. In Section 4.2, the examples are extensions of examples that appear in other CCPS books. The subsequent sections describe other examples submitted by the committee from various industries.

There are two aspects of probability of ignition (POI) prediction that can be misinterpreted, and it is worthwhile to review these in advance of the case studies. This is discussed next.

4.1.1 "Reality" vs. Predictions

Several of the examples in Sections 4.3 and later represent actual events that happened in the chemical process industry. While the algorithms in this book should be expected to roughly replicate the expected ignition probabilities for an *aggregate* of incidents, it should be remembered that they will never be absolutely "correct" for any *single* incident. After all, the probability of a single release igniting is defined by discrete values—either it *did* ignite (POI = 1) or it *did not* (POI = 0). In contrast, the POI algorithms will almost always give an answer somewhere *between* 0 and 1.

The accuracy of the tool described in this book can be debated on similar grounds, since the algorithms are based on finite data of "typical" situations, and the tool could potentially be applied to situations not foreseen. However, a subjective view based on initial testing is that the methods in this book have roughly the following expected ranges:

Level 1 Analysis—80% of "true" ignition probabilities expected to be within a factor of 1.75 of the value predicted in this book.

Level 2 Analysis—80% of true ignition probabilities expected to be within a factor of 1.5 of the value predicted in this book.

Level 3 Analysis—80% of true ignition probabilities expected to be within a factor of 1.25 of the value predicted in this book.

The expectations above do not describe "confidence limits" in the classic sense, but rather are in recognition that there will always be cases that are exceptions to the rules. There is no substitute for field checks to determine the applicability of the methods in this book to a specific and possibly unusual set of conditions. See also the discussion in Section 4.1.3.

4.1.2 "Conservatism"—Does It Exist?

As a general principle, risk analyses of whatever type should strive for accuracy, but where accuracy is not absolute, the analyses should err on the side of conservatism (overstating the risk). This principle could be incorporated in two ways in this book:

1. *Overall Conservatism*—The final result should err on the side of conservatism.

2. *Conservatism with Respect to Expected Level of Accuracy*—Since more effort is put into a Level 3 analysis than a Level 1 analysis, it would be preferred that a benefit in terms of reduction in conservatism would be associated with the extra effort needed for the Level 3 analysis.

While the methods described in this book strive to achieve these ideas *overall*, it is important to remember that there will be many individual cases in which they do not. It is also important to recognize that in a model that contains several inputs, introducing conservatism into each input could result in a final outcome that is conservative by over an order of magnitude.

Following are some brief illustrations of the complexities of introducing conservatism into an ignition probability model.

Case 1—Immediate Ignition vs. Delayed Ignition/Explosion

Consider an approach to the development of the algorithms that result in events being "conservatively" predicted to have a higher probability of immediate ignition than might be expected based on historical data. For that matter, one could intend to be ultraconservative and design algorithms that would predict that the event ignited at the source in *all* cases. This may well be conservative in many cases, particularly if there are buildings/populations in the immediate vicinity that are vulnerable to fire.

However, the conservative assumption of 100% immediate ignition precludes the possibility of a delayed ignition. The consequences of a delayed ignition could be much more severe than an immediate ignition, particularly if the outcome is an explosion. Thus, eliminating the possibility of an explosion by developing a onservative algorithm for probability of immediate ignition defeats the intent of erring on the side of conservatism.

Case 2—Releases of Flammables That Are Also Toxic

One can also imagine a case in which a released material is both toxic and flammable. In some cases, the fire/explosion effects, and toxic products of combustion, are less severe than the toxic effects of the source chemical. The fire/explosion thus precludes the source chemical toxic event from happening some fraction of the time. In such cases, utilizing a conservative ignition probability model would be conservative with respect to fire and explosion outcomes but nonconservative with respect to toxic outcomes. An illustration of this is provided in Section 4.5.2.

Case 3—Conservatisms in Level 1 vs. Level 2 vs. Level 3

As a general philosophy, it is preferred for a Level 2 or 3 analysis to "reward" the extra effort required by producing a result that is more accurate than those resulting from the lower level(s) of analysis and reduces the conservatism intended to be in the methods at the lower level(s). However, this cannot always be the case, and one can readily devise examples of this point.

Consider the case of a butane release in a refining unit. The Level 1 analysis will typically give a result where the overall probability of ignition is about 0.3, and history has shown that on the whole most flammable releases do not ignite. However, if the release occurs just a short distance from a fired heater, ignition may be almost certain. As a result, the Level 2 and 3 analyses will give higher POI predictions than the Level 1 analysis. In this particular case, the Level 1 approach is nonconservative, even though in the entire universe of release events, Level 1 would give conservative results relative to Levels 2 and 3 more often than not.

Such are the conundrums associated with trying to balance "conservatism" and "accuracy" in a methodology that deals with probabilities rather than absolutes.

4.1.3   Cases Where the Model May Not Be Appropriate or the Results Misinterpreted

The models described in this book are designed to work in typical onshore process plant applications—that is, where a flammable release is at some distance from the relevant ignition sources and the vapor cloud and/or liquid spill is expected to contact one or more ignition sources. There are extreme cases at either end of "some distance" in which the model can be expected to break down. These are:

- *Ignition Source Is Extremely Close to the Point of Release*—An obvious example of this would be the release of process fluid from the tubes inside a fired heater.
- *Ignition Source Is Extremely Remote from the Point of Release*—There is a universe of ignition sources available; only those that can be reached by the flammable cloud are relevant. A similar outcome can arise when the release is very small; see discussions in Sections 1.2.7 and 2.1.1.

In the extreme cases above, the actual probability of delayed ignition (PODI) may approach 1 or 0, but the model will not know this because it is not a consequence (spatial) model. In these cases the analyst should either enter the obvious answer for the probability or utilize a dispersion model and wind direction information to determine the probability that the ignition source(s) would be in contact with the vapor cloud. The probability that the vapor cloud can reach an ignition source and the probability of the wind direction blowing the cloud toward an ignition source can be captured in an event tree along with the PODI estimate to develop final PODI results. See also the discussions in Section 2.5.

4.1.4 Summary of Worked Examples

Following are several illustrations of the algorithms proposed in this book. Aside from providing mathematical replication of the methods, the examples provide useful lessons to be learned regarding applications and limitations of the methods. Note there will always be some disconnect between the examples in the book and the software available with this book, since the two do not have precisely the same capabilities, and the software will be edited over time. A summary of the examples is provided in Table 4.1.

Table 4.1. Summary of worked examples

Section	Title	Themes
4.2.1	Vapor Cloud Explosion Hazard Assessment of a Storage Site	Continuation of example from other CCPS book. Treatment of vehicles as point or line ignition sources. Illustrates combining multiple ignition sources.
4.2.2	Open Field Release of Propane	Continuation of example from other CCPS book. Comparison of results of Level 1, 2, and 3 approaches.
4.2.3	Release from Pipeline	Continuation of example from other CCPS book. Pipeline example.
4.3.1	Ethylene Tubing Failure	Replication of an actual event. Discussion of treatment of release plume orientation.
4.3.2	Benzene Pipe Rupture	Replication of an actual event using Level 1 analysis.
4.3.3	Spill from Methyl Ethyl Ketone Tank	Comparison of Level 1 and Level 2 analyses.
4.3.4	Indoor Puncture of MEK Tote	Illustration of Level 3 methods for indoor releases.
4.3.5	Elevated Release	Discussion of thoughtful assessment of ignition sources when the release is elevated above most expected ignition sources.
4.4.1	Gasoline Release from a Sight Glass	Discussion of the complications associated with vapor generation after a release has ended and potential for ignition sources to be deactivated during the release event.
4.4.2	Overfilling a Gasoline Storage Tank	Discussion of the complications associated with vapor generation after a release has ended. Comparison with actual event.
4.4.3	Overfilling a Propane Bullet	Discusses need for user to incorporate dispersion model results into an ignition assessment. Comparison with actual event.

Table 4.1 (continued)

Section	Title	Themes
4.4.4	Hydrogen Release from a Sight Glass	Hydrogen case illustration. Also discussion about treatment of toxic component of ignitable stream to prevent nonconservative results.
4.5.1	Indoor Acid Spill—Ventilation Model	Discusses whether to model a volatile liquid release as a liquid or as a vapor. Complex indoor/outdoor situation.
4.5.2	Release of Ammonia	Comparison of Level 1 and Level 2 event trees for a toxic material that is also marginally flammable.
4.6.1	Release of Gas from an Offshore Platform Separator	Demonstration of potential application to offshore facility.
4.6.2	Dust Ignition	Extension of methods to dust example that in principle should be well outside the intended scope of this book.
4.7.1	Ignition by Hot Surfaces	Compares ignition by hot surfaces for two chemicals with much different AIT values.
4.7.2	Release Prevention	General discussion; not an illustration.
4.7.3	Duration of Exposure	Illustration of effect of event duration; discussion of other methods for preventing the ignition.
4.7.4	Benefit of Improved Ventilation of Indoor Releases	Continuation of "Indoor Acid Spill" example.

Note that during some of the calculations that follow three significant digits are used. The reader should not infer that there is such a degree of accuracy in the methods; rather, the excess digits are retained to avoid rounding errors during the calculations. The final values are reported to two significant digits, although that is probably overstating the accuracy as well.

4.2   WORKED EXAMPLES (BASED ON OTHER CCPS BOOKS)

4.2.1   Vapor Cloud Explosion Hazard Assessment of a Storage Site

4.2.1.1   Introduction to Example

This is a continuation of an example from page 218 of the CCPS book "Guidelines for Vapor Cloud Explosion, Pressure Vessel Burst, BLEVE and Flash Fire Hazards" (CCPS, 2010). The area layout is provided in Figure 4.1, which shows a 50-m-diameter butane storage tank as well as three propane storage spheres.

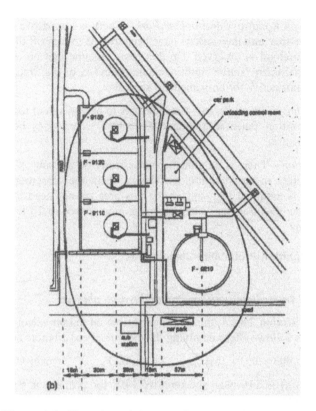

Figure 4.1. Plot plan of the tank farm (CCPS, 2010).

The example was originally used to illustrate different methods for quantifying the magnitude of an explosion involving a release of 20,000 kg from one of the propane spheres. In this book, the same example is used to quantify the probability that such a release will ignite. For the purposes of this illustration, a Level 2 analysis is performed.

4.2.1.2   The Event Description

The event to be modeled was described in the earlier book as a release from a 0.1-m- (4-inch-) diameter leak in the unloading line from propane sphere F-9120 at ambient temperature of 293K (68 °F) and pressure of 8 bars (116 psig). The resulting flammable range of the dispersion is depicted by the oval shape shown in Figure 4.1.

4.2.1.3   Potential Ignition Sources

For the purposes of this example, it is assumed that the following ignition sources are present:

Traffic on Roadways and in Car Park—There is normally a minor amount of traffic in the area, and the workers have been trained to turn off their ignition when a flammable cloud is observed. On this basis, assume that on average there is a single car that is an "active" ignition source about ½ of the time, but this ignition source remains active for only about 30 seconds.

Unloading Control Room—Consider this to be equivalent to "office space" in terms of ignition potential (assume the control room does not operate under positive pressure).

Substation—There is no guidance specifically on the subject of substations as ignition sources in the literature, and details about the substation are not known, although it is assumed that voltages in this area do not need to be exceedingly high. Assume that it is equivalent as an ignition source to 20 feet of high-power electrical line.

4.2.1.4   Ignition Probability Calculations

Level 2 Immediate Ignition Probability Calculations

As per Section 2.8.1.1, the "static" contribution to immediate ignition can be expressed as follows after combining the individual contributors and modifiers:

$$POII_{static} = 0.003 \times P^{1/3} \times \{MIE_{reported} \times (10,000/P_{liquid})^{0.25} \times \exp[0.0044(60 - T)]\}^{-0.6}$$

P ($=P_{liquid}$) is 116 psig, the reported MIE for propane is 0.25 mJ, and T is 68 °F. Therefore,

$$POII_{static} = 0.003 \times (116)^{1/3} \times \{(0.25) \times (10,000/116)^{0.25}$$

$$\times \exp[0.0044(60 - 68)]\}^{-0.6}$$

$$= 0.018$$

Section 2.8.1.2 describes the contribution of autoignition. Since the temperature of propane is well below its AIT of 842 °F, there is no contribution of autoignition to the POII. The predicted overall POII is therefore 0.018.

Level 2 Delayed Ignition Probability Calculations

There are three identified ignition sources, each of which must be calculated individually using the algorithms described in Section 2.8.2.

Traffic on Roadways and in Car Park

The baseline PODI is based on ignition source strength and event duration. The source strength, as listed in Table 2.2, is

$$S = 1 - 0.7^V$$

It was previously assumed that, on average, 0.5 cars would be "active" and only for 30 seconds (t = 0.5 minutes). Then,

$$S = 1 - 0.7^{0.5} = 0.163 \text{ (when active)}$$

The baseline PODI is calculated as per Section 2.8.2.2:

$$PODI_{S/D} = 1 - [(1 - S^2) \times e^{-St}]$$

$$PODI_{S/D} = 1 - [(1 - 0.163^2) \times e^{-0.163 \times 0.5}] = 0.103$$

The first factor to be applied to the baseline PODI is the "Magnitude of Release" multiplier. This may be estimated in one of two ways, as described in Section 2.8.2.3:

$$M_{MAG\_Amount\ Released\ (liquid)} = (Amount\ Released/5,000)^{0.3}$$

$$M_{MAG\_Hole\ Diameter\ (liquid)} = (Hole\ Diameter)^{0.6}$$

For this example, the first of the equations above gives a value of ~1.9; the second a value of ~2.3. Therefore a value of 2.1 is chosen for M_{MAG}.

The second multiplier to the PODI is based on the MIE of the material being released. Per Section 2.8.2.4,

$$M_{MAT} = 0.5 - 1.7 \log(MIE), \text{ or}$$

$$M_{MAT} = 0.5 - 1.7 \log(0.25) = 1.52$$

The third factor to be applied is based on the release temperature vs. normal boiling point (or flash point), as described in Section 2.8.2.5. In this case, the boiling point of propane is well below the release temperature, and so a multiplier of 1 is used. Similarly, the release is outdoors, so the indoor/outdoor multiplier described in Section 2.8.2.6 is 1.

The combined PODI for the first ignition source is then

$$PODI_{Level\ 2} = PODI_{S/D} \times M_{MAG} \times M_{MAT} \times M_T \times M_{IN/OUT}$$

$$PODI_{Level\ 2} = 0.103 \times 2.1 \times 1.52 \times 1 \times 1 = 0.32$$

The other two ignition source probabilities are calculated next.

Unloading Control Room

As before, the baseline PODI is based on ignition source strength and event duration. The source strength for an "office space," as listed in Table 2.2, is S = 0.05. In this case, the effective duration is the entire 4 minutes of the release, so that

$$PODI_{S/D} = 1 - [(1 - 0.05^2) \times e^{-0.05 \times 4}] = 0.183$$

The PODI modifiers remain the same as for the previous ignition source, and therefore the combined PODI for the second ignition source is

$$PODI_{Level\ 2} = 0.183 \times 2.1 \times 1.52 \times 1 \times 1 = 0.57$$

Substation

In this case, Table 2.2 and the "20-foot equivalent high-power line" assumption results in the source strength of 0.02, which, for 4 minutes of exposure, results in

$$PODI_{S/D} = 1 - [(1 - 0.02^2) \times e^{-0.02 \times 4}] = 0.077$$

The PODI modifiers remain the same as for the previous ignition source, and therefore the combined PODI for the second ignition source is

$$PODI_{Level\ 2} = 0.077 \times 2.1 \times 1.52 \times 1 \times 1 = 0.24$$

Combination of Ignition Sources

The final PODI from the three ignition sources above is not additive, which can be seen by considering two perspectives. In one way of viewing the problem, if ignition probabilities were additive, then there is the potential for the sum of the probabilities to be greater than 1, which is not logical. From a different perspective, the cloud can only be ignited once; once the ignition has taken place, the other two ignition sources are irrelevant.

For this reason, the probabilities of ignition must be combined using Boolean addition as follows:

$$PODI = [0.32 + 0.57 - (0.32 \times 0.57)] + 0.24$$

$$- \{[0.32 + 0.57 - (0.32 \times 0.57)] \times 0.24\}$$

$$= 0.78$$

Alternatively, the probabilities of *non*-ignition can be *multiplied*, as described in Section 2.10, then subtracted from 1. In this case, the result is as follows:

$$PODI_{combined} = 1 - \prod(PODI_{individual\ ignition\ sources})$$

$$= 1 - (1 - 0.32)(1 - 0.57)(1 - 0.24) = 0.78$$

Final Level 2 Ignition Probability Calculations

POII was estimated above to be 0.017. Since immediate ignition precludes the possibility of a delayed ignition, the probability of delayed ignition is

$$PODI_{final} = PODI_{calc}(1 - POII) = 0.78(1 - 0.018) = 0.77$$

This PODI includes both fire and explosion outcomes.

Discussion—Motor vehicle ignition sources can be handled either as point source(s) or as a line source, as per Table 2.2. If there is more than one "active" vehicle, it is probably more convenient to use the "line-source" approach. Note also that the initial version of the software associated with this book allows for only a single common event duration, so that manual calculation of multiple ignition sources is necessary.

4.2.2 Open Field Release of Propane

4.2.2.1 Introduction to Example

This is a continuation of an example from page 92 of the CCPS book "Guidelines for Vapor Cloud Explosion, Pressure Vessel Burst, BLEVE and Flash Fire Hazards" (CCPS, 2010). This example involves a "massive" release of propane into an open field, but in this case the release is instantaneous and the "cloud assumes a flat, circular shape as it spreads," reaching an undefined ignition source when it has dimensions of 1 meter depth and 100 meters diameter (Figure 4.2).

The original example assumed that an explosion was not possible because turbulence-inducing obstructions were not present. In this book, the same example is used to quantify the probability that such a release will ignite. For the purposes of comparison, Level 1, 2, and 3 analyses are performed.

4.2.2.2 The Event Description

The details of the discharge are not known but are inferred later by the description of the resulting cloud.

4.2.2.3 Potential Ignition Sources

For the purposes of this example, it is assumed that the following ignition source is present:

General Ignition Source—This is an area that can be described as "remote outdoor storage area." The event is described as "instantaneous," but the original example describes wind at a modest 2 m/s, and so it is assumed here that the cloud would persist in the flammable range for 3 minutes.

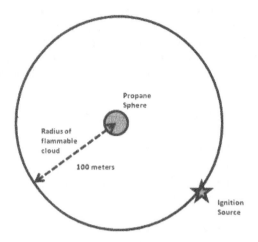

Figure 4.2. Open field release of propane.

4.2.2.4   Ignition Probability Calculations

The Level 1, Level 2, and Level 3 calculations are described next.

Level 1 Immediate Ignition Probability Calculations

As per Section 2.7.1.1, the "static" contribution to immediate ignition for Level 1 is simply assumed to be 0.05. Since the temperature of propane is well below its AIT, there is no contribution of autoignition to the POII. The predicted overall POII is therefore 0.05.

Level 1 Delayed Ignition Probability Calculations

The release is outdoors. Therefore, as per Section 2.7.2, since the chemical is on the standard pick list, the PODI is calculated automatically as per Eq. 2-17:

$$PODI = 0.15 - 0.25 \log MIE$$

$$PODI = 0.15 - 0.25 \log(0.25) = 0.30$$

Level 1 POI Summary

POII was estimated above to be 0.05. Since immediate ignition precludes the possibility of a delayed ignition, the probability of delayed ignition is

$$PODI_{final} = PODI_{calc} (1 - POII) = 0.30 (1 - 0.05) = 0.28$$

This PODI includes both fire and explosion outcomes.

Level 2 Immediate Ignition Probability Calculations

As per Section 2.8.1.1, the "static" contribution to immediate ignition can be expressed as follows after combining the individual contributors and modifiers. It is assumed that the propane is initially stored at 68 °F and 116 psig:

$$POII_{static} = 0.003 \times P^{1/3} \times \{MIE_{reported} \times (10,000/P_{liquid})^{0.25} \times exp[0.0044(60 - T)]\}^{-0.6}$$

P ($=P_{liquid}$) is 116 psig, the reported MIE for propane is 0.25 mJ, and T is 68 °F. Therefore,

$$POII_{static} = 0.003 \times (116)^{1/3} \times \{(0.25) \times (10,000/116)^{0.25} \times exp[0.0044(60 - 68)]\}^{-0.6}$$

$$= 0.018$$

Section 2.8.1.2 describes the contribution of autoignition. Since the temperature of propane is well below its AIT of 932 °F, there is no contribution of autoignition to the POII. The predicted overall POII is therefore 0.018.

Level 2 Delayed Ignition Probability Calculations

There is a single area ignition source defined. As per Table 2.2, this "remote outdoor storage area" has a "strength" of S = 0.025.

The baseline PODI is calculated as per Section 2.8.2.2:

$$PODI_{S/D} = 1 - [(1 - S^2) \times e^{-St}]$$

$$PODI_{S/D} = 1 - [(1 - 0.025^2) \times e^{-0.025 \times 3}] = 0.073$$

The first factor to be applied to the baseline PODI is the "Magnitude of Release" multiplier. Based on the description of the cloud dimensions, and a concentration of 10 vol% in the cloud (from the original example description), the amount released is about 3,500 pounds. Therefore, from Section 2.8.2.3,

$$M_{MAG\_Amount\ Released\ (liquid)} = (Amount\ Released/5,000)^{0.3}$$

$$M_{MAG\_Amount\ Released\ (liquid)} = (3,500/5,000)^{0.3} = 0.899$$

The second multiplier to the PODI is based on the MIE of the material being released. Per Section 2.8.2.4,

$$M_{MAT} = 0.5 - 1.7 \log(MIE), \text{ or}$$

$$M_{MAT} = 0.5 - 1.7 \log(0.25) = 1.52$$

The third factor to be applied is based on the release temperature vs. normal boiling point, as described in Section 2.8.2.5. In this case, the boiling point of propane is well below the release temperature, and so a multiplier of 1 is used.

Similarly, the release is outdoors, so the indoor/outdoor multiplier described in Section 2.8.2.6 is 1.

The combined PODI is then

$$PODI_{Level\ 2} = PODI_{S/D} \times M_{MAG} \times M_{MAT} \times M_T \times M_{IN/OUT}$$

$$PODI_{Level\ 2} = 0.073 \times 0.899 \times 1.52 \times 1 \times 1 = 0.10$$

Level 2 POI Summary

POII was estimated above to be 0.018. Since immediate ignition precludes the possibility of a delayed ignition, the probability of delayed ignition is

$$PODI_{final} = PODI_{calc}\ (1 - POII) = 0.10(1 - 0.018) = 0.10$$

This PODI includes both fire and explosion outcomes.

Level 3 Immediate Ignition Probability Calculations

As per Section 2.9.1, the only difference between a Level 2 and Level 3 POII "static" contribution to immediate ignition is in accounting for the temperature of the material after release rather than its temperature in the process. In this case, the temperature of the release drops to its normal boiling point of –44 °F. Using the same relationship as was used for the Level 2 analysis, but with –44 °F, yields the following:

$$POII_{static} = 0.003 \times P^{1/3} \times \{MIE_{reported} \times (10,000/P_{liquid})^{0.25}\ exp[0.0044(60 - T)]\}^{-0.6}$$

P $(=P_{liquid})$ is 116 psig, the reported MIE for propane is 0.25 mJ, and T is –44 °F. Therefore,

$$POII_{static} = 0.003 \times (116)^{1/3} \times \{(0.25)\ (10,000/116)^{0.25}\ exp[0.0044(60 - (-44))]\}^{-0.6}$$

$$= 0.013$$

Section 2.8.1.2 describes the contribution of autoignition. Since the temperature of propane is well below its AIT, there is no contribution of autoignition to the POII. The predicted overall POII is therefore 0.013.

Level 3 Delayed Ignition Probability Calculations

The enhancements that are possible at Level 3 include the following:

[Per Section 2.9.2.1] Modification of "S" to account for how well ignition sources are managed. In this case, it will be assumed that the management of ignition sources is "typical" and so no modification is used compared to the Level 2 analysis.

Other new modifications exist, but only for inside releases. Therefore, the remainder of the analysis is the same as for Level 2:

$$PODI_{Level\ 2} = PODI_{S/D} \times M_{MAG} \times M_{MAT} \times M_T \times M_{IN/OUT}$$

$$PODI_{Level\ 2} = 0.073 \times 0.899 \times 1.52 \times 1 \times 1 = 0.10$$

Level 3 POI Summary

POII was estimated above to be 0.013. Since immediate ignition precludes the possibility of a delayed ignition, the probability of delayed ignition is

$$PODI_{final} = PODI_{calc} (1 - POII) = 0.10 (1 - 0.013) = 0.10$$

This PODI includes both fire and explosion outcomes.

Comparison of Levels 1, 2, and 3 POI Results

It is interesting to note the comparison in results for the three levels:

Level	POII	PODI
1	0.05	0.30
2	0.018	0.10
3	0.013	0.10

In this case, Levels 2 and 3 remove some of the conservatism that is intended to be built into the Level 1 analysis, as desired.

4.2.3 Release from Pipeline

4.2.3.1 Introduction to Example

This is a continuation of an example that starts on the bottom of page 240 of the CCPS book "Guidelines for Chemical Process Quantitative Risk Analysis" (CCPS, 1999). The example was originally used to demonstrate the calculations for determining the radiant heat flux from a jet fire that might occur following the pipeline release. In this book, the same example is used to quantify the probability that such a release will ignite. For the purposes of this illustration, a Level 2 analysis is performed.

4.2.3.2   The Event Description

The event to be modeled was described in the earlier book (CCPS, 1999) as a release from a 25-mm- (1-inch-) diameter leak in a methane pipeline (Figure 4.3). The pipeline is at a pressure of 100 bars gauge (1450 psig) and is assumed to be

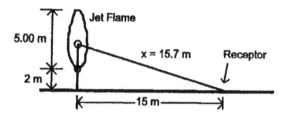

FIGURE 2.83. Geometry for Example 2.31: Radiant flux from a jet fire.

**Figure 4.3. Illustration of pipeline release
(Jet fire outcome from CCPS, 1999).**

operating at the ambient temperature of 298K (77 °F). The release is oriented vertically. The inference of the term "pipeline" is that it is remote from process equipment, but since the thermal radiation from a jet fire is of interest, it is assumed that the pipeline passes through a sparsely populated residential area.

4.2.3.3   Potential Ignition Sources

For the purposes of this example, it is assumed that the following ignition sources are present:

Residential Population—Based on the description above, it is assumed that there is one household within range of the release and that there are normally two occupants present (Figure 4.4).

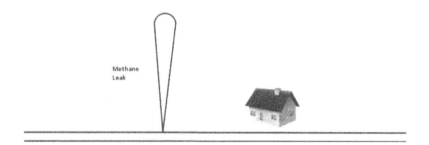

Figure 4.4. Release from pipeline.

Note that this is a 1-inch hole in what is presumed to be a much larger diameter pipeline, and so it is assumed that the initial release can be sustained for an extended period. However, it will also be assumed that people present would evacuate quickly, such that they (and the major ignition sources in the vicinity of their house) have been removed after 3 minutes.

4.2.3.4   Ignition Probability Calculations

Level 2 Immediate Ignition Probability Calculations

As per Section 2.8.1.1, the "static" contribution to immediate ignition can be expressed as follows for vapor releases:

$$POII_{static} = 0.003 \times P^{1/3} \times MIE^{-0.6}$$

P is 1450 psig, the reported MIE for methane is 0.21 mJ, and T is 77 °F. Therefore,

$$POII_{static} = 0.003 \times (1450)^{1/3} \times (0.21)^{-0.6} = 0.091$$

Section 2.8.1.2 describes the contribution of autoignition. Since the pipeline temperature is well below methane's AIT of ~1,000 °F, there is no contribution of autoignition to the POII. The predicted overall POII is therefore 0.091.

Level 2 Delayed Ignition Probability Calculations

The baseline PODI is based on ignition source strength and event duration. The source strength, as listed in Table 2.2, is

$$S = 1 - 0.99^N$$

It was assumed that, on average, two people would be present, so that

$$S = 1 - 0.99^2 = 0.0199$$

The baseline PODI is calculated as per Section 2.8.2.2, given an assumed ignition source duration of 3 minutes:

$$PODI_{S/D} = 1 - [(1 - S^2) \times e^{-St}]$$
$$PODI_{S/D} = 1 - [(1 - 0.0199^2) \times e^{-0.0199 \times 3}] = 0.0583$$

The first factor to be applied to the baseline PODI is the "Magnitude of Release" multiplier. The hole size is known, so per Section 2.8.2.3,

$$M_{MAG\_Hole\ Diameter\ (vapor)} = (Hole\ Diameter) = 1$$

The second multiplier to the PODI is based on the MIE of the material being released. Per Section 2.8.2.4,

$$M_{MAT} = 0.5 - 1.7\ log(MIE)$$

or

$$M_{MAT} = 0.5 - 1.7 \log(0.21) = 1.65$$

The third factor to be applied is based on the release temperature vs. normal boiling point, as described in Section 2.8.2.5. In this case, the boiling point of methane is well below the release temperature, and so a multiplier of 1 is used. Similarly, the release is outdoors, so the indoor/outdoor multiplier described in Section 2.8.2.6 is 1.

The combined PODI for the first ignition source is then

$$PODI_{Level\ 2} = PODI_{S/D} \times M_{MAG} \times M_{MAT} \times M_T \times M_{IN/OUT}$$
$$PODI_{Level\ 2} = 0.0583 \times 1 \times 1.65 \times 1 \times 1 = 0.0964$$

Final Level 2 Ignition Probability Calculations

POII was estimated above to be 0.091. Since immediate ignition precludes the possibility of a delayed ignition, the probability of delayed ignition is

$$PODI_{final} = PODI_{calc}\ (1 - POII) = 0.0964(1 - 0.091) = 0.088$$

This PODI includes both fire and explosion outcomes.

Discussion of Results

The analysis above is based on "conventional" ignition sources. However, it should be remembered that there are two other very likely causes of ignition in pipeline events—(1) from external contact from an excavator or other equipment that was also the cause of the release and (2) ignition from jet-induced particle formation. Such events should be evaluated for both frequency of release and ignition probability outside the context of this ignition probability model and managed using preventive controls.

4.3   WORKED EXAMPLES (CHEMICAL AND PETROCHEMICAL PLANTS)

4.3.1   Ethylene Tubing Failure

4.3.1.1   The Event Description

900 pounds (400 kg) of ethylene gas is released over a 2-minute period when a ½-inch tubing line fails due to fatigue. The operating pressure in the line is 1925 psig (135 bars) and the temperature is 225 °F (107 °C). The release occurs at the

edge of a process unit having typical ignition sources. A Level 2 analysis is to be performed.

4.3.1.2   Potential Ignition Sources

The following ignition sources are present:

Process Area—This is characterized as a "medium-density process."

4.3.1.3   Ignition Probability Calculations

Level 2 Immediate Ignition Probability Calculations

As per Section 2.8.1.1, the "static" contribution to immediate ignition can be expressed as follows after combining the applicable individual contributors and modifier:

$$POII_{static} = 0.003 \times P^{1/3} \times \{MIE_{reported} \times \exp[0.0044(60 - 225)]\}^{-0.6}$$

P is 1925 psig, the reported MIE for ethylene is about 0.084 mJ, and T is 225 °F. Therefore,

$$POII_{static} = 0.003 \times (1925)^{1/3} \times \{(0.084) \times \exp[0.0044(60 - 225)]\}^{-0.6}$$

$$= 0.255$$

Section 2.8.1.2 describes the contribution of autoignition. Since the temperature of the ethylene is well below its AIT, there is no contribution of autoignition to the POII. The predicted overall POII is therefore 0.255.

Level 2 Delayed Ignition Probability Calculations

There is a single area ignition source to consider. The source strength, as listed in Table 2.2, is $S = 0.15$. The baseline PODI is calculated as per Section 2.8.2.2:

$$PODI_{S/D} = 1 - [(1 - S^2) \times e^{-St}]$$

$$PODI_{S/D} = 1 - [(1 - 0.15^2) \times e^{-0.15 \times 2}] = 0.276$$

The first factor to be applied to the baseline PODI is the "Magnitude of Release" multiplier. This can be estimated using either the diameter of the failed pipe or the amount released, as described in Section 2.8.2.3:

$$M_{MAG\_Hole\ Diameter\ (vapor)} = (\text{Hole Diameter}) = 0.5$$

$$M_{MAG\_Amount\ Released\ (vapor)} = (\text{Amount Released}/1,000)^{0.5} = (900/1,000)^{0.5} = 0.95$$

Assume an average of the two of about 0.7.

The second multiplier to the PODI is based on the MIE of the material being released. Per Section 2.8.2.4,

$$M_{MAT} = 0.5 - 1.7 \log(MIE), \text{ or}$$

$$M_{MAT} = 0.5 - 1.7 \log(0.084) = 2.33$$

The third factor to be applied is based on the release temperature vs. normal boiling (or flash) point, as described in Section 2.8.2.5, but this does not apply to vapor releases. Also, the release is outdoors, so the indoor/outdoor multiplier described in Section 2.8.2.6 is 1.

The combined PODI for the first ignition source is then

$$PODI_{Level\ 2} = PODI_{S/D} \times M_{MAG} \times M_{MAT} \times M_T \times M_{IN/OUT}$$

$$PODI_{Level\ 2} = 0.276 \times 0.7 \times 2.33 \times 1 \times 1 = 0.45$$

Final Level 2 Ignition Probability Calculations

POII was estimated above to be 0.255. Since immediate ignition precludes the possibility of a delayed ignition, the probability of delayed ignition is

$$PODI_{final} = PODI_{calc}\ (1 - POII) = 0.45\ (1 - 0.255) = 0.335$$

This PODI includes both fire and explosion outcomes.

Discussion of Results

This example illustrated an anecdotal report of a release that did not ignite in the actual event. The algorithms in this book predict that the release would most likely not ignite, but with a significant probability of ignition.

Since the release is at the edge of the unit, it may be appropriate to multiply the ignition probability by the fraction of events that would be oriented into the unit. This fraction may be governed by the following: (a) the angular "view" of the source to the rest of the unit, (b) the wind direction, and (c) the momentum (pressure) of the release.

4.3.2   Benzene Pipe Rupture

4.3.2.1   The Event Description

In the event that an 8-inch line ruptures, benzene at 270 psig and 100 °F would be released into a "typical" process unit. A Level 1 analysis is to be performed for input to a LOPA study.

4.3.2.2 Ignition Probability Calculations

Level 1 Immediate Ignition Probability Calculations As per Section 2.7.1.1, the "static" contribution to immediate ignition for Level 1 is simply assumed to be 0.05. Since the temperature of benzene is well below its AIT, there is no contribution of autoignition to the POII. The predicted overall POII is therefore 0.05.

Level 1 Delayed Ignition Probability Calculations

The release is outdoors. Therefore, as per Section 2.7.2, since the chemical is on the standard pick list, the PODI is calculated automatically as per Eq. 2-17:

$$PODI = 0.15 - 0.25 \log MIE$$

$$PODI = 0.15 - 0.25 \log(0.2) = 0.325$$

Final Level 1 Ignition Probability Calculations

POII was estimated above to be 0.05. Since immediate ignition precludes the possibility of a delayed ignition, the probability of delayed ignition is

$$PODI_{final} = PODI_{calc} (1 - POII) = 0.325(1 - 0.05) = 0.31$$

This PODI includes both fire and explosion outcomes.

Discussion of Results

This event actually occurred, and no ignition resulted. The algorithms in this book suggest about one chance in three of ignition.

4.3.3 Spill from Methyl Ethyl Ketone Tank

4.3.3.1 The Event Description

Methyl ethyl ketone (MEK) is stored at 20 °C under a slight nitrogen pad in a 7,500-gallon vertical storage tank. The scenario is a tank leak or discharge line leak/failure that would empty the tank contents to the dike and spill pit over a short period (10 minutes to a few hours). Because of the leak location, it is not expected that it could be safely stopped.

There is a good chance that a tank truck would be unloading in the adjacent area about 50 feet from the tank, and the dike drainage would bring a draining pool of MEK liquid within 25 feet of the tank truck. The tank truck is not running during unloading; however, it would be running during positioning and the exhaust would be hot for about 10 minutes after tank truck positioning. Thus, a high-temperature ignition source could exist 10% of the time during unloading, with tank truck unloading occurring 25% of the time during the day. The wind, if from the west (P = 0.3), would blow the MEK vapor cloud over the tank truck area with the plume distance to the LFL being 30 feet and the plume being 20 feet wide for a release rate of 15 kg/s (the lowest release rate case of interest). For a 150-kg/s (~ "catastrophic") release rate, the LFL cloud is still only 40 feet long and 40 feet wide. There could be other trucks or cars in the area but they would be more than 50 feet from the dike. The key question is, what would the POI be, given that an LFL cloud blows toward the hot engine exhaust?

4.3.3.2   Potential Ignition Sources

There are some aspects of the problem that can be solved using the tools in this book, and some that cannot. For the time being, it will be assumed that the cloud can reach the ignition source (that the wind direction and truck presence permit this). The conditional probabilities for these will be discussed later.

Figure 4.5. Spill from MEK tank.

There is just one ignition source—the hot truck engine. The calculations will be performed at both Level 1 and Level 2.

4.3.3.3   Ignition Probability Calculations

Immediate Ignition Probability Calculations

Level 1—As per Section 2.7.1, the value of POII is set to 0.05, since there is no contribution from autoignition at the ambient storage temperature.

Level 2—As per Section 2.8.1.1, the contributions to immediate ignition include a static contribution that is dependent on the tank pressure and the effective MIE of the material being released:

$$POII_{static} = 0.003 \times P^{1/3} \times \{MIE_{reported} \times (10,000/P_{liquid})^{0.25} \times \exp[0.0044(60 - T)]\}^{-0.6}$$

P ($=P_{liquid}$) is considered to be 5 psig, to account for liquid head in the tank. The MIE for MEK is 0.53 mJ. T is 68 °F.

Therefore,

$$POII_{static} = 0.003 \times (5)^{1/3} \times \{(0.53) \times (10,000/5)^{0.25} \times \exp[0.0044(60 - 68)]\}^{-0.6}$$

$$= 0.00245$$

Section 2.8.1.2 describes the contribution of autoignition. Since the temperature of this mixture is well below its AIT, there is no contribution of autoignition to the POII. The predicted overall POII is therefore 0.00245.

Delayed Ignition Probability Calculations

Level 1—As per Section 2.7.2, since the chemical is on the standard pick list, the PODI is calculated automatically as per Eq. 2-17:

$$PODI = 0.15 - 0.25 \log MIE$$

$$PODI = 0.15 - 0.25 \log(0.53) = 0.22$$

Level 2—The strength S can be calculated one of two ways, either by assuming that the truck is a "motor vehicle" as per Table 2.2 or by treating the truck as a "hot surface" as per the discussion following Table 2.2. S for each case is

$$S_{motor\ vehicle} = 0.3$$

$$S_{hot\ surface} = 0.5 + 0.0025[1,000 - 941 - 100(1)] = 0.40$$

[In the calculation above, the temperature of the engine is assumed to be 1000 °F, the velocity of the gas flowing over the engine is assumed to be 1 m/s, and the autoignition temperature of MEK is 941 °F.] Select an average value of 0.35.

The baseline PODI is calculated as per Section 2.8.2.2, given the original statement that the truck engine could remain hot for 10 minutes:

$$PODI_{S/D} = 1 - [(1 - S^2) \times e^{-St}]$$

$$PODI_{S/D} = 1 - [(1 - 0.35^2) \times e^{-0.35 \times 10}] = 0.973$$

The first factor to be applied to the baseline PODI is the "Magnitude of Release" multiplier. This can be estimated using the amount released for the two release cases described earlier, then applying the weight-based relationship in Section 2.8.2.3:

Amount released at 15 kg/s = (15 kg/s)(2.2 lb/kg)(600 s) = 19,800 lb

and

$$M_{MAG\_Amount\ Released\ (liquid)} = (Amount\ Released/5,000)^{0.3} = 1.51$$

Amount released at 150 kg/s = (150 kg/s) (2.2 lb/kg)(600 s) = 198,000 lb. This amount is greater than the tank capacity of 7500 gallons (~50,000 lb), so it is assumed that the release amount is limited to 50,000 lb, and the event duration is limited to 2.5 minutes. This in turn changes the earlier baseline calculation for PODI to the following:

$$PODI_{S/D(150kg/s)} = 1 - [(1 - 0.35^2) \times e^{-0.35 \times 2.5}] = 0.634$$

$$M_{MAG\_Amount\ Released\ (liquid)} = (50,000/5,000)^{0.3} = 2.00$$

The second multiplier to the PODI is based on the MIE of the material being released. Per Section 2.8.2.4,

$$M_{MAT} = 0.5 - 1.7\ log(MIE),\ or$$

$$M_{MAT} = 0.5 - 1.7\ log(0.53) = 0.97$$

The third factor to be applied is based on the release temperature vs. normal boiling point, as described in Section 2.8.2.5:

$$M_T = 1 - (NBP - T)/230,\ or$$

$$M_T = 1 - (175 - 68)/230 = 0.535$$

The last multiplier is for indoor vs. outdoor releases, which is set to 1 for this outdoor release. Then,

$$PODI_{Level\ 2} = PODI_{S/D} \times M_{MAG} \times M_{MAT} \times M_T \times M_{IN/OUT}$$

$$PODI_{Level\ 2,\ 15kg/s} = 0.973 \times 1.51 \times 0.97 \times 0.535 \times 1 = 0.762$$

$$PODI_{Level\ 2,\ 150kg/s} = 0.634 \times 2.00 \times 0.97 \times 0.535 \times 1 = 0.658$$

Final Ignition Probability Calculations

Level 1—POII and PODI were estimated above to be 0.05 and 0.22, respectively. Since immediate ignition precludes the possibility of a delayed ignition, the probability of delayed ignition is

$$PODI_{final} = PODI_{calc}\ (1 - POII) = 0.22(1 - 0.05) = 0.21$$

This PODI includes both fire and explosion outcomes.

Level 2—POII was estimated above to be 0.00245. PODI was estimated to be 0.762 and 0.658, respectively, for the 15- and 150-kg/s release cases. Since immediate ignition precludes the possibility of a delayed ignition, the probability of delayed ignition is

$$PODI_{final,\ 15kg/s} = PODI_{calc}\ (1 - POII) = 0.762(1 - 0.00245) = 0.76$$

$$PODI_{final,\ 150kg/s} = PODI_{calc}\ (1 - POII) = 0.658(1 - 0.00245) = 0.66$$

This PODI includes both fire and explosion outcomes.

Discussion of Results

Here is a case where the Level 2 analysis gives higher delayed ignition probabilities than the Level 1 analysis, thus illustrating the points made in Section 4.2.1 about conservatism in probability calculations.

One factor that could be incorporated into the analysis outside of the algorithms in this book is the probability of the wind blowing in the unfavorable wind direction (toward the hot engine). If the wind blows toward the hot engine 25% of the time, the PODI would be multiplied by 0.25, giving a PODI of 0.165 (close to Level 1 results).

4.3.4 Indoor Puncture of MEK Tote

4.3.4.1 The Event Description

This evaluates an MEK tote spill due to fork truck puncture inside a large building at 20 °C with 220 gallons of MEK spilling out over 10 minutes. The MEK spill footprint was about 50 feet long and 30 feet wide and about 1.5 inches deep at the low curb. In this case, the flammable limits were verified with explosimeters as being about 20 feet beyond the edge of the spill area. The area is Class I, Division 2. The air changes per hour were greater than 10 with a room exhaust fan activated. There were people within the LFL zone for a limited time. What would the POI be? (This was an actual incident with no ignition but much excitement).

A Level 3 analysis is to be performed so that the effect of ventilation (supplemental model from Appendix B) can be taken into account for the indoor ignition probability calculation.

Further Information—The room in which the spill occurs has dimensions of about 80 feet × 125 feet × 15 feet tall.

4.3.4.2 Potential Ignition Sources

The following ignition sources are present:

Indoor Process Area—This is characterized as a "medium-density process." While there were people present that might themselves be characterized as ignition sources, it was assumed that they simply represented the normal population of a medium-density process area.

4.3.4.3 Ignition Probability Calculations

Level 3 Immediate Ignition Probability Calculations

As per Section 2.9.1, the only difference between a Level 2 and Level 3 POII "static" contribution to immediate ignition is in accounting for the temperature of the material after release rather than its temperature in the process. In this case, the temperatures are the same, so that

$$POII_{static} = 0.003 \times P^{1/3} \times \{MIE_{reported} \times (10{,}000/P_{liquid})^{0.25} \times \exp[0.0044(60 - T)]\}^{-0.6}$$

P (=P_{liquid}) is considered to be 1 psig, to account for liquid head in the tote. The MIE for MEK is 0.53 mJ, and T is 68 °F. Therefore,

$$POII_{static} = 0.003 \times (1)^{1/3} \times \{(0.53) \times (10,000/1)^{0.25} \times exp[0.0044(60 - 68)]\}^{-0.6}$$

$$= 0.0154$$

Section 2.8.1.2 describes the contribution of autoignition. Since the temperature of this mixture is well below its AIT, there is no contribution of autoignition to the POII. The predicted overall POII is therefore 0.0154.

Level 3 Delayed Ignition Probability Calculations

The enhancements that are possible at Level 3 relative to Level 2 include the following:

[Per Section 2.9.2.1] Modification of S to account for how well ignition sources are managed. In this case, it will be assumed that the management of ignition sources as Class I, Division 2 is "typical" for this kind of application, and so no modification is used compared to the Level 2 analysis.

[Per Appendix B] Inside release modifier that accounts for ventilation.

The remainder of the analysis is the same as for Level 2, but with the modifications above.

Indoor Process Area—This has been described as a "medium-density process" area. The source strength, as listed in Table 2.2, is $S = 0.15$. The baseline PODI is calculated as per Section 2.8.2.2:

$$PODI_{S/D} = 1 - [(1 - S^2) \times e^{-St}]$$

$$PODI_{S/D} = 1 - [(1 - 0.15^2) \times e^{-0.15 \times 10}] = 0.782$$

The first factor to be applied to the baseline PODI is the "Magnitude of Release" multiplier. This can be estimated using the amount released (220 gallons ~ 1475 lb), as described in Section 2.8.2.3:

$$M_{MAG\_Amount\ Released\ (liquid)} = (Amount\ Released/5,000)^{0.3} = 0.693$$

The second multiplier to the PODI is based on the MIE of the material being released. Per Section 2.8.2.4,

$$M_{MAT} = 0.5 - 1.7 \log(MIE), or$$

$$M_{MAT} = 0.5 - 1.7 \log(0.53) = 0.969$$

The third factor to be applied is based on the release temperature (or flash point) vs. normal boiling point, as described in Section 2.8.2.5:

$$M_T = 1 - (NBP - T)/230, or$$

$$M_T = 1 - (175 - 68)/230 = 0.535$$

The last multiplier accounts for an indoor release, and in Level 3, the ventilation may be taken into account as per Appendix B given the following inputs:

Building volume (V) = 80 ft × 125 ft × 15 ft = 150,000 ft^3

Effective ventilation rate (EVR) is known to be greater than 10 air changes per hour, but 10 ACH is assumed to be conservative.

Draft Direction—It is assumed that the draft is not specifically designed to draw vapors away from likely ignition sources inside the building ($B_{vdd} = 1$).

As per Appendix B, the indoor multiplier M_V is then

$$M_V = 1.5 \times B_{es} \times B_{vr} \times B_{vdd}$$

$$B_{es} = (V/150,000)^{-1/3} = (150,000/150,000)^{-1/3} = 1.00$$

$$B_{vr} = (EVR/2)^{-1/2} = (10/2)^{-1/2} = 0.447$$

$$M_V = 1.5 \times 1.00 \times 0.447 \times 1 = 0.671$$

Lastly,

$$PODI_{Level\ 3inside} = PODI_{S/D} \times M_{MAG} \times M_{MAT} \times M_T \times M_V$$

$$PODI_{Level\ 3inside} = 0.782 \times 0.693 \times 0.969 \times 0.535 \times 0.671 = 0.189$$

Final Level 3 Ignition Probability Calculations

POII was estimated above to be 0.0154. Since immediate ignition precludes the possibility of a delayed ignition, the probability of delayed ignition is

$$PODI_{final} = PODI_{calc} (1 - POII) = 0.189 (1 - 0.0154) = 0.19$$

This PODI includes both fire and explosion outcomes.

Discussion of Results

This example could be extended to determine the benefit of having even higher ventilation rates than those that currently exist, as illustrated in the example in Section 4.7.4. Alternatively, performing a comparison with lower ventilation rates might be used as a justification for keeping the current system if people complain either about the cost of operating the current system or about any discomfort that arises from having high ventilation (e.g., room being cold in the winter).

4.3.5   Elevated Release

4.3.5.1   The Event Description

In this case a runaway reaction involving ethylene oxide (EO) is resulting in a vessel overpressure that is being vented to atmosphere through a relief valve. It has been determined that the relief device is adequately sized and that the vessel is not filled so high that a two-phase release occurs. The PSV releases at 5 m above grade with a 90-degree bend resulting in a flammable EO vapor cloud that travels 45 m downwind into the plant process area which has structures that are 15 m high and 50 m wide and are located only 15 m from the EO vent line. Thus, 30 m of the EO cloud is in the process structure, with a width of 40 m and a cloud height of 8 m at least 3 m above grade. The cloud basically impacts the second level of the process structure's three levels. This is a congested area with 9600 m^3 of flammable EO vapor in the congested structure. A dispersion model is used to determine if the EO vapor cloud can fill the second level zone.

The EO release temperature is 153 °F (67 °C) and the release pressure is 76 psig (5.3 barg). The process area is Class 1, Division 2 with an operator usually present. The release rate is 46 kg/s for 530 s to mostly vent 25,000 kg of EO.

4.3.5.2   Potential Ignition Sources

The following ignition sources are present:

Process Area—In a general sense this area would be characterized as a "medium-density process." However there are very few sources of ignition in the elevated portion of the structure into which the release is predicted to discharge. On that basis, the analyst might choose to artificially treat the area as a "low-density process." The analysis will proceed on that basis, with discussion following. A Level 2 analysis will be performed.

4.3.5.3   Ignition Probability Calculations

Level 2 Immediate Ignition Probability Calculations

Per Section 2.8.1.1, the "static" contribution to immediate ignition can be expressed as follows after combining the applicable individual contributors and modifier:

$$POII_{static} = 0.003 \times P^{1/3} \times \{MIE_{reported} \times exp[0.0044(60 - T)]\}^{-0.6}$$

P is 76 psig, the reported MIE for ethylene oxide is about 0.065 mJ, and T is 153 °F.

Therefore,

$$POII_{static} = 0.003 \times (76)^{1/3} \times \{(0.065) \times exp[0.0044(60 - 153)]\}^{-0.6}$$

$$= 0.084$$

Section 2.8.1.2 describes the contribution of autoignition. Since the temperature of the EO is well below its AIT of 804 °F (429 °C), there is no contribution of autoignition to the POII. The predicted overall POII is therefore 0.084.

Level 2 Delayed Ignition Probability Calculations

There is a single area ignition source to consider; for the reasons described earlier, it will be treated as a "low-density process." The source strength, as listed in Table 2.2, is $S = 0.1$, and the event duration is 530 seconds (8.8 minutes). The baseline PODI is calculated as per Section 2.8.2.2:

$$PODI_{S/D} = 1 - [(1 - S^2) \times e^{-St}]$$

$$PODI_{S/D} = 1 - [(1 - 0.1^2) \times e^{-0.1 \times 8.8}] = 0.589$$

The first factor to be applied to the baseline PODI is the "Magnitude of Release" multiplier. This can be estimated using the amount released (25,000 kg, 55,000 lb), as described in Section 2.8.2.3:

$$M_{MAG\_Amount\ Released\ (vapor)} = (Amount\ Released/1,000)^{0.5} = (55,000/1,000)^{0.5} = 7.42$$

This value exceeds the limit on $M_{MAG\_Amount\ Released}$ of 2, so it is reset to 2.

The second multiplier to the PODI is based on the MIE of the material being released. Per Section 2.8.2.4,

$$M_{MAT} = 0.5 - 1.7 \log(MIE), or$$

$$M_{MAT} = 0.5 - 1.7 \log(0.065) = 2.52$$

The third factor to be applied is based on the release temperature vs. normal boiling (or flash) point, as described in Section 2.8.2.5, but this does not apply to vapor releases. Also, the release is outdoors, so the indoor/outdoor multiplier described in Section 2.8.2.6 is 1.

The combined PODI is then

$$PODI_{Level\ 2} = PODI_{S/D} \times M_{MAG} \times M_{MAT} \times M_T \times M_{IN/OUT}$$

$$PODI_{Level\ 2} = 0.589 \times 2 \times 2.52 \times 1 \times 1$$

Since this result is greater than 1, delayed ignition is considered to be virtually assured.

Level 2 Probability of Explosion Calculations

Explosion probabilities are generally outside the scope of the algorithms in this book to address since they are dependent on congestion/confinement factors that are necessarily addressed in a consequence model. And for that reason, explosion probability models are discussed in detail only in Appendix B and, in isolated test cases, in the examples in this chapter. However, in this case the material being released has a high fundamental burning velocity, and is being discharged into a moderately congested space. For those reasons one might reasonably conclude that an explosion is physically possible without the aid of a consequence model. Therefore, this example will proceed using the approach for explosion probability modeling described in Appendix B.

As described in Appendix B, there are three factors that are used as multipliers onto a "base" value of 0.3 for probability of explosion given delayed ignition (POEGDI):

$$POEGDI_{Level\ 2} = 0.3 \times M_{CHEM} \times M_{MAGE} \times M_{IN/OUT}$$

These terms are calculated as follows:

M_{CHEM}—M_{CHEM} is a function of the chemical's "reactivity," or propensity to ignite explosively as measured by its fundamental burning velocity. Ethylene is a "high-reactivity" material, and so M_{CHEM} has a value of 2.

M_{MAGE}—M_{MAGE} is similar to M_{MAG} but is not considered as strong an influence as for PODI. Per Appendix B,

$$M_{MAGE} = (PODI\ M_{MAG})^{0.5}$$
$$M_{MAGE} = (2)^{0.5} = 1.41$$

$M_{IN/OUT}$—$M_{IN/OUT}$ is the multiplier used if a release is indoors and is calculated as for PODI. For this outdoor release, the value is 1.

Thus the overall Level 2 POEGDI is

$$POEGDI_{Level\ 2} = 0.3 \times 2 \times 1.41 \times 1 = 0.85$$

Final Level 2 Ignition Probability Calculations

POII was estimated above to be 0.084. Since immediate ignition precludes the possibility of a delayed ignition, the probability of delayed ignition is

$$PODI_{final} = PODI_{calc}\ (1 - POII) = 1\ (1 - 0.084) = 0.916$$

This PODI includes both fire and explosion outcomes. The fraction of delayed ignitions that result in explosions is calculated above as 0.85.

Therefore the probabilities of each outcome are:

POII = 0.084

PODI resulting in an explosion = 0.916 (0.85) = 0.78

PODI resulting in a fire only = 0.916 (0.15) = 0.14

Discussion of Results

This example illustrated a case where the standard definition of an area ignition source may be replaced with one that better reflects the known presence (or absence) of ignition sources in the area. Even so, the model predicted that ignition was virtually assured.

The model is probably conservative in doing so; however, there may be reason to retain conservatism in this situation. In this event the jet is predicted to discharge horizontally and remain elevated until reaching its flammable limit. The discharge is into a congested space. Whereas the discharge model predicts an uninterrupted release, in fact the congestion may well result in impingement of the jet. This impingement may result in a broader-than-predicted plume and/or slumping of the cloud to lower elevations once its momentum has been disturbed.

Lastly, it may be imprudent to assume that there is no toxic outcome because this model predicted 100% ignition. There may be reasons (e.g., wind direction) that a release does not contact an ignition source, and this may be incorporated in an event tree. But even in the absence of such effects, the user should assume a low, but nonzero, probability of nonignition so that the toxic branch of the outcome event tree is not eliminated.

4.4 WORKED EXAMPLES (OIL REFINERIES)

4.4.1 Gasoline Release from a Sight Glass

4.4.1.1 The Event Description

A sight glass with a ¾-inch pipe connection at column C-101 fails, releasing gasoline at 3 bars (45 psig) and 120 °C (248 °F). The release rate is determined to be about 4 kg/s (530 lb/min). It is estimated that the distance from the source to the cloud LFL (the circle) is 40 m (130 feet). The column is located in an outside high-density process area surrounded by Roads G, H, 19, and 20 (Figure 4.6).

Road 20 is the main pathway to the other parts of the refinery and has very busy vehicle traffic (>100 cars/day). The traffic on Roads 19, G, and H is moderate, with about 15 cars/day on 24 hours average. There are also two furnaces close to the release location. Assume that operator response to the release results in C-101 being depressurized sufficiently that emergency response teams can approach and isolate the sight glass in 15 minutes. A Level 2 analysis will be performed.

4.4.1.2   Potential Ignition Sources

The following ignition sources are present:

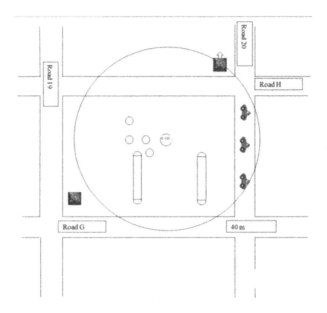

Figure 4.6. Location of sight glass failure and environs.

Traffic on Road 20—Road 20 is the main pathway to the other parts of the refinery and has very busy vehicle traffic (>100 cars/day). If the path through the flammable cloud on Road 20 is about 40 m (130 feet), and the speed limit is 15 miles/hour (25 km/h), then the average fractional vehicle presence in a flammable cloud on Road 20 is

(120 cars/day)(40 m/car)(day/24 h)(h/15 miles)(mile/1609 m) = 0.0082

Traffic on Roads 19, G, and H—These roads average 15 cars/day each. If the path through the flammable cloud is the same as Road 20 (including Road 19, to be conservative) and the speed limit is 15 miles/hour (25 km/h), then the average fractional vehicle presence in a flammable cloud on these roads is

(45 cars/day)(40 m/car)(day/24 h)(h/15 miles)(mile/1609 m) = 0.0031

Process Area—This is characterized as a "high-density process."

Fired Heaters—Two fired heaters were mentioned as being close to the release source. While these heaters could be considered a subset of the high-density process area, the ignition source "strength" assigned to fired heaters is higher than for a high-density process, so the heater values are used and the process source is considered minor by comparison to avoid overlap counting of ignition sources.

4.4.1.3 Ignition Probability Calculations

Level 2 Immediate Ignition Probability Calculations

Per Section 2.8.1.1, the "static" contribution to immediate ignition can be expressed as follows after combining the individual contributors and modifiers:

$$POII_{static} = 0.003 \times P^{1/3} \times \{MIE_{reported} \times (10,000/P_{liquid})^{0.25} \times \exp[0.0044(60 - T)]\}^{-0.6}$$

P ($=P_{liquid}$) is 45 psig, the reported MIE for gasoline varies from 0.23 to 0.8 mJ (assume 0.4 mJ), and T is 248F. Therefore,

$$POII_{static} = 0.003 \times (45)^{1/3} \times \{(0.4) \times (10,000/45)^{0.25} \times \exp[0.0044(60 - 248)]\}^{-0.6}$$

$$= 0.013$$

Section 2.8.1.2 describes the contribution of autoignition. Since the temperature of gasoline is well below its AIT, there is no contribution of autoignition to the POII. The predicted overall POII is therefore 0.013.

Level 2 Delayed Ignition Probability Calculations

There are four ignition sources to consider, each of which must be calculated individually using the algorithms described in Section 2.8.2.

Traffic

The baseline PODI is based on ignition source strength and event duration. The source strength, as listed in Table 2.2, is

$$S = 1 - 0.7^V$$

Considering first only the traffic on Road 20, it was previously calculated that on average 0.0082 cars would be in the flammable cloud. Given the stated distance and vehicle speed in the cloud, they would be in the cloud for about 5.8 seconds (t = 0.1 minutes). Then,

$$S = 1 - 0.7^{0.0082} = 0.00292$$

The baseline PODI is calculated as per Section 2.8.2.2:

$$PODI_{S/D} = 1 - [(1 - S^2) \times e^{-St}]$$

$$PODI_{S/D} = 1 - [(1-0.00292^2) \times e^{-0.00292 \times 0.1}] = 0.00030$$

The first factor to be applied to the baseline PODI is the "Magnitude of Release" multiplier. This is estimated using the diameter of the failed pipe, as described in Section 2.8.2.3:

$$M_{MAG\_Hole\ Diameter\ (liquid)} = (Hole\ Diameter)^{0.6} = 0.75^{0.6} = 0.84$$

The second multiplier to the PODI is based on the MIE of the material being released. Per Section 2.8.2.4,

$$M_{MAT} = 0.5 - 1.7 \log(MIE), \text{ or}$$

$$M_{MAT} = 0.5 - 1.7 \log (0.4) = 1.18$$

The third factor to be applied is based on the release temperature vs. normal boiling point, as described in Section 2.8.2.5. In this case, the gasoline contains a range of boiling point materials, some with boiling points above the release temperature, some below. Here it will be assumed that the release temperature is also the normal boiling point, and so a multiplier of 1 is used. Similarly, the release is outdoors, so the indoor/outdoor multiplier described in Section 2.6.2.6 is 1.

The combined PODI for the first ignition source is then

$$PODI_{Level\ 2} = PODI_{S/D} \times M_{MAG} \times M_{MAT} \times M_T \times M_{IN/OUT}$$

$$PODI_{Level\ 2} = 0.00030 \times 0.84 \times 1.18 \times 1 \times 1 = 0.00030$$

The other vehicle ignition sources can be calculated using the same methods to get an additional $PODI_{Level\ 2} = 0.00011$.

Fired Heaters (Each)

As before, the baseline PODI is based on ignition source strength and event duration. The source strength for a fired heater, as listed in Table 2.2, is $S = 0.9$. In this case the effective duration is the entire 15 minutes of the release; however, the duration is artificially limited to 10 minutes for the reasons described in Section 2.8.2.2. Even so,

$$PODI_{S/D} = 1 - [(1 - 0.9^2) \times e^{-0.9 \times 10}] \sim 1$$

Therefore, in this case the other ignition sources are trivial by comparison, and delayed ignition is virtually assured.

Final Level 2 Ignition Probability Calculations

POII was estimated above to be 0.013. Since immediate ignition precludes the possibility of a delayed ignition, the probability of delayed ignition is

$$PODI_{final} = PODI_{calc} (1 - POII) = 1 (1 - 0.013) = 0.99$$

This PODI includes both fire and explosion outcomes.

Discussion of Results

The analysis above could be considered conservative with respect to several implied assumptions. The release was stated as having an effective duration of 15 minutes, and the subsequent calculations were based on this duration *and* the implied assumption that the ignition sources are active throughout this period. In fact, one would expect that the fired heaters would have been shut down, and traffic stopped, prior to the release being isolated. Thus the analysis could be refined further to account for these factors.

A potential nonconservatism in the analysis is the implied assumption that the exposure stops once the leak has been isolated. In fact, there is likely to be a pool of gasoline from which vapors would continue to emanate. The implied assumption in this case is that emergency responders are able to apply fire-fighting foam or other means to prevent vapor evolution from the pool—at the same time as the leak is being isolated.

4.4.2 Overfilling a Gasoline Storage Tank

4.4.2.1 The Event Description
A gasoline tank is being filled remotely at a rate of 5400 lb/min (41 kg/s). There is a failure of the tank level control system and the tank begins to overfill. It takes 10 minutes before the operator closes the tank inlet valve. The surroundings of the storage tank are a typical tank farm area, with no other point ignition sources present. A Level 2 analysis will be performed.

4.4.2.2 Potential Ignition Sources
The only ignition sources present are from the equipment in the tank farm itself, which can reasonably be described as a "remote outdoor storage area" as per Table 2.2.

4.4.2.3 Ignition Probability Calculations

Level 2 Immediate Ignition Probability Calculations

As per Section 2.8.1.1, the "static" contribution to immediate ignition can be expressed as follows after combining the individual contributors and modifiers:

$$POII_{static} = 0.003 \times P^{1/3} \times \{MIE_{reported} \times (10,000/P_{liquid})^{0.25} \times exp[0.0044(60 - T)]\}^{-0.6}$$

For the purposes of this example it is assumed that the pressure in the tank P ($=P_{liquid}$) is 5 psig. The reported MIE for gasoline varies from 0.23 to 0.8 mJ (assume 0.4 mJ), and T is slightly above ambient (30 °C/86 °F). Therefore,

$$POII_{static} = 0.003 \times (5)^{1/3} \times \{(0.4) \times (10,000/5)^{0.25} \times exp[0.0044(60 - 86)]\}^{-0.6}$$

$$= 0.0030$$

Section 2.8.1.2 describes the contribution of autoignition. Since the temperature of the gasoline is well below its AIT, there is no contribution of autoignition to the POII. The predicted overall POII is therefore 0.0030.

Level 2 Delayed Ignition Probability Calculations

The single area ignition source has an ignition source strength S of 0.025, as per Table 2.2. The duration of the overfill is stated as being 10 minutes. In actuality, the effective duration of the flammable cloud is likely to be longer, as vapors evolve from the pool of gasoline. However, the duration is limited to 10 minutes for the reasons described in Section 2.8.2.2. The baseline PODI is then calculated as per Section 2.8.2.2:

$$PODI_{S/D} = 1 - [(1 - S^2) \times e^{-St}]$$

$$PODI_{S/D} = 1 - [(1 - 0.025^2) \times e^{-0.025 \times 10}] = 0.222$$

The first factor to be applied to the baseline PODI is the "Magnitude of Release" multiplier. This is estimated using the total quantity released—41 kg/s for 10 minutes = 24,600 kg (54,200 lb). From Section 2.8.2.3,

$$M_{MAG\_Amount\ Released\ (liquid)} = (54,200/5,000)^{0.3} = 2.04$$

The second multiplier to the PODI is based on the MIE of the material being released. Per Section 2.8.2.4,

$$M_{MAT} = 0.5 - 1.7 \log(MIE), \text{ or}$$

$$M_{MAT} = 0.5 - 1.7 \log(0.4) = 1.18$$

The third factor to be applied is based on the release temperature vs. normal boiling point, as described in Section 2.8.2.5. In this case, the gasoline is stored at a much lower temperature than its normal boiling point, and so a multiplier of 1 is used. Similarly, the release is outdoors, so the indoor/outdoor multiplier described in Section 2.8.2.6 is 1.

The combined PODI for the first ignition source is then

$$PODI_{Level\ 2} = PODI_{S/D} \times M_{MAG} \times M_{MAT} \times M_T \times M_{IN/OUT}$$

$$PODI_{Level\ 2-10min} = 0.222 \times 2.04 \times 1.18 \times 1 \times 1 = 0.54$$

Final Level 2 Ignition Probability Calculations

POII was estimated above to be 0.003. Since immediate ignition precludes the possibility of a delayed ignition, the probability of delayed ignition is

$$PODI_{final} = PODI_{calc} (1 - POII) = 0.54(1 - 0.003) = 0.54$$

This PODI includes both fire and explosion outcomes.

Discussion of Results

The analysis above could be considered conservative with respect to the effective duration. While the overfill event may continue for 10 minutes before being stopped, there is some initial period of time in which the gasoline vapors will not have been able to reach any ignition source that would be typical of a tank farm area.

Actual Event: Spill with Later Foam Application

One anecdote reported for this book described an actual event in which foam was applied after 1,700 barrels (440,000 lb; 200,000 kg) had been spilled. No ignition occurred.

4.4.3 Overfilling a Propane Bullet

4.4.3.1 The Event Description

A large propane bullet is being filled but is overfilled, resulting in propane being released to atmosphere through a pressure relief valve at the fill rate of 8,000 lb/min (60 kg/s). The event continues until a total of 100,000 lb (45,000 kg) is released. The bullet is located in a relatively remote storage area. The relief valve has a set pressure of 150 psig (10 bars), with the discharge directed vertically at an elevation of 30 feet (10 meters) above grade. A Level 2 analysis will be performed.

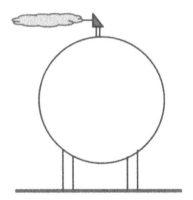

Figure 4.7. Relief from a propane bullet.

4.4.3.2   Potential Ignition Sources

The only ignition sources that are present are from the equipment in the bullet area and surrounding tank farm, which can be described as a "remote outdoor storage area" as per Table 2.2. However, a dispersion analysis indicates that the released material does not approach ground level in the flammable range. Since there are no credible elevated ignition sources in the flammable vapor cloud, only immediate ignition is considered.

4.4.3.3   Ignition Probability Calculations

Level 2 Immediate Ignition Probability Calculations

As per Section 2.8.1.1, the "static" contribution to immediate ignition can be expressed as follows after combining the individual contributors and modifiers:

$$POII_{static} = 0.003 \times P^{1/3} \times \{MIE_{reported} \times (10,000/P_{liquid})^{0.25} \exp[0.0044(60 - T)]\}^{-0.6}$$

The relief pressure is 150 psig, the reported MIE for propane is 0.25 mJ, and T is ambient ($\sim$ 70 °F/21 °C). Therefore,

$$POII_{static} = 0.003 \times (150)^{1/3} \times \{(0.25) \times (10,000/150)^{0.25} \exp[0.0044(60 - 70)]\}^{-0.6}$$

$$= 0.020$$

Section 2.8.1.2 describes the contribution of autoignition.   Since the temperature of the propane is well below its AIT, there is no contribution of autoignition to the POII.   The predicted overall POII, and the overall POI, is therefore 0.020.

Discussion of Results

The analysis above illustrates an important feature about this book—that the tools in this book do not include dispersion models, and the user is responsible for determining if a release could actually contact a given ignition source and, if it could, the probability that contact would take place. Note that the software that incorporates the methods in this book does not have such logic built into it either; it will simply report a probability for delayed ignition that in some cases may be highly unlikely. It is therefore incumbent on the user to assess the situation thoughtfully and to conclude whether the delayed ignition results should be ignored.

There are perhaps exceptions to the assumption made in this example of having "no credible elevated ignition sources" in a remote tank farm area. For example, if a release could occur as the result of severe weather in the area, then lightning from the same storm could potentially ignite the release. This constitutes a "common cause" combination of events that is beyond the capability of the tools in this book. However, the analyst should consider whether there is a possibility of an ignition source being created coincident with the release of interest and account for it using methods other than those described in this book.

Incidentally, a very similar event to the one in this example was reported to this book committee. No ignition was observed, which appears to be consistent with the low probability of ignition predicted by these methods.

4.4.4 Hydrogen Release from a Sight Glass

4.4.4.1 The Event Description

A gauge glass on a hydrotreater separator fails. The release passes through a ¾-inch valve at the glass, releasing a recycle hydrogen stream consisting of 90 vol% hydrogen, 8% methane, and 2% ethane and heavier gases. The operating conditions at the separator are 1850 psig and 135 °F.

Assume that operator response to the release results in the separator being depressurized sufficiently that emergency response teams can approach and isolate the sight glass in 10 minutes. The hydrotreater is conservatively considered to be a "high-density process" area. A Level 2 analysis is being performed.

4.4.4.2 Ignition Probability Calculations

Level 2 Immediate Ignition Probability Calculations

As per Section 2.8.1.1, the "static" contribution to immediate ignition can be expressed as follows after combining the individual contributors and modifiers:

$$POII_{static} = 0.003 \times P^{1/3} \times \{MIE \times \exp[0.0044(60 - T)]\}^{-0.6}$$

P is 1,850 psig and T is 135 °F. The release is a mixture of chemicals having the following MIEs:

Hydrogen — 0.016 mJ

Methane — 0.21 mJ

Ethane — 0.27 mJ

Assuming a proportional mixing rule, the average MIE = 0.016(0.9) + 0.21(0.08) + 0.27(0.02) = 0.0366 mJ. Then

$$POII_{static} = 0.003 \times (1850)^{1/3} \times \{(0.0366) \times \exp[0.0044(60 - 135)]\}^{-0.6}$$

$$= 0.254$$

Section 2.8.1.2 describes the contribution of autoignition. Since the temperature of this stream is well below its AIT, there is no contribution of autoignition to the POII. The predicted overall POII is therefore 0.254.

Level 2 Delayed Ignition Probability Calculations

The "high-density process" area ignition source has a value of strength S of 0.25, as per Table 2.2. Then, given a 10-minute duration, the baseline PODI is calculated per Section 2.8.2.2:

$$PODI_{S/D} = 1 - [(1 - S^2) \times e^{-St}]$$

$$PODI_{S/D} = 1 - [(1 - 0.25^2) \times e^{-0.25 \times 10}] = 0.923$$

The first factor to be applied to the baseline PODI is the "Magnitude of Release" multiplier. This is estimated using the diameter of the failed pipe, as described in Section 2.8.2.3:

$$M_{MAG\_Hole\ Diameter\ (vapor)} = (Hole\ Diameter)^1 = 0.75^1 = 0.75$$

The second multiplier to the PODI is based on the MIE of the material being released. Per Section 2.8.2.4,

$$M_{MAT} = 0.5 - 1.7 \log(MIE), \text{ or}$$

$$M_{MAT} = 0.5 - 1.7 \log(0.0366) = 2.94$$

The third factor to be applied is based on the release temperature vs. normal boiling point, as described in Section 2.8.2.5, but for a vapor release this is set to 1. Also, the release is outdoors, so the indoor/outdoor multiplier described in Section 2.8.2.6 is 1.

The combined PODI for the first ignition source is then

$$PODI_{Level\ 2} = PODI_{S/D} \times M_{MAG} \times M_{MAT} \times M_T \times M_{IN/OUT}$$

$$PODI_{Level\ 2} = 0.923 \times 0.75 \times 2.94 \times 1 \times 1 = 2.04$$

Since this value is greater than 1, it should be assumed that the release will ignite in all cases.

Final Level 2 Ignition Probability Calculations

POII was estimated above to be 0.254. Since immediate ignition precludes the possibility of a delayed ignition, the probability of delayed ignition is

$$PODI_{final} = PODI_{calc}(1 - POII) = 1(1 - 0.254) = 0.746$$

This PODI includes both fire and explosion outcomes.

Discussion of Results

The analysis suggests that the event will always ignite. For the reasons described in the discussion of hydrogen in Section 1.7.1, this might be overstating the situation. However, absent any specific knowledge about the local environment in the aftermath of the release (whether or not the release is inhibited by obstructions at the point of release), and given the other components that are present in the mixture, it is prudent to accept the predictions as is.

However, if there had been a toxic component (e.g., hydrogen sulfide) to this stream, it would have been imprudent to assume that there is no toxic outcome because this model predicted ignition. As with an earlier example, the user should assume a low, but nonzero, probability of nonignition.

4.5   WORKED EXAMPLES (UNUSUAL CASES)

In the example cases in this section, there are some unusual aspect(s) to the analysis. Either:

- Analysis requires use of a more speculative/complicated model, or
- Material being released is both flammable and toxic.

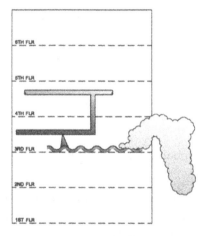

Figure 4.8. Indoor spill that migrates outdoors.

4.5.1   Indoor Acid Spill—Ventilation Model

4.5.1.1   The Event Description

A 1-inch drain valve is left open in a piping system and causes a release during startup of an operation. The material in the pipe is 95% acetic acid/5% water at 260 psig and 205 °C. The spill occurs on the third floor of a six-story building with solid concrete floors (Figure 4.8). The processing building has a significant number of instruments, electrical connections, and rotating pieces of equipment. The drainage is adequate to remove the spilled material, but the flashed acid spreads throughout that floor of the structure and on to the north of the structure where there is a rail spur (traffic 1% of the time) and vehicular traffic (traffic 5% of the time). No fired equipment is in the immediate vicinity. A Level 3 analysis is to be performed so that the effect of ventilation can be taken into account for the indoor ignition probability calculation.

Further Information—The floor on which the spill occurs has dimensions of about 30 feet × 50 feet × 20 feet tall, and the minimum ventilation rate is three air changes per hour in the winter. For the purposes of this assessment, it is assumed that vapors in the flammable range can reach the outside ignition sources. The release is assumed to be isolated remotely after a period of 2 minutes, but the analysis will assume that significant vapors can be generated from the pool for another minute after that.

4.5.1.2   Potential Ignition Sources

The following ignition sources are present:

Indoor Process Area—This is characterized as a "medium-density process area."

Diesel Train—Present only 1% of the time

Motor Vehicles—Present 5% of the time

These sources need to be assessed individually.

4.5.1.3   Ignition Probability Calculations

Level 3 Immediate Ignition Probability Calculations

As per Section 2.9.1, the only difference between a Level 2 and Level 3 POII "static" contribution to immediate ignition is in accounting for the temperature of the material after release rather than its temperature in the process. In this case, the temperature of the release reduces to its normal boiling point of 239 °F (115 °C). Using the same relationship defined for a Level 2 analysis but with a temperature of 239 °F yields the following:

$$POII_{static} = 0.003 \times P^{1/3} \times \{MIE_{reported} \times (10,000/P_{liquid})^{0.25} \times exp[0.0044(60 - T)]\}^{-0.6}$$

P ($=P_{liquid}$) is 260 psig. MIE data for acetic acid solutions is not readily available, but a value of 7 mJ was selected based on tests published by Garland (2010), and T is 239 °F. Therefore,

$$POII_{static} = 0.003 \times (260)^{1/3} \times \{(7) \times (10{,}000/260)^{0.25} \times \exp[0.0044(60 - 239)]\}^{-0.6}$$

$$= 0.0055$$

Section 2.8.1.2 describes the contribution of autoignition. Since the temperature of this mixture is well below its AIT, there is no contribution of autoignition to the POII. The predicted overall POII is therefore 0.0055.

Level 3 Delayed Ignition Probability Calculations

These will need to be calculated separately for each ignition source. The enhancements that are possible at Level 3 relative to Level 2 include the following:

[Per Section 2.9.2.1] Modification of S to account for how well ignition sources are managed. In this case, it will be assumed that the management of ignition sources is "typical" and so no modification is used compared to the Level 2 analysis.

[Per Appendix B] Inside release modifier that accounts for ventilation.

The remainder of the analysis is the same as for Level 2, but with the modifications above. Each ignition source will now be reviewed.

Indoor Process Area

This has been described as a "medium-density process area." The source strength, as listed in Table 2.2, is S = 0.15. The baseline PODI is calculated as per Section 2.8.2.2:

$$PODI_{S/D} = 1 - [(1 - S^2) \times e^{-St}]$$

$$PODI_{S/D} = 1 - [(1 - 0.15^2) \times e^{-0.15 \times 3}] = 0.377$$

The first factor to be applied to the baseline PODI is the "Magnitude of Release" multiplier. This can be estimated using the diameter of the open line, as described in Section 2.8.2.3:

$$M_{MAG\_Hole\ Diameter\ (liquid)} = (Hole\ Diameter)^{0.6} = 1$$

The second multiplier to the PODI is based on the MIE of the material being released. Per Section 2.8.2.4,

$$M_{MAT} = 0.5 - 1.7 \log(MIE), \text{ or}$$

$$M_{MAT} = 0.5 - 1.7 \log(7) = -0.93$$

This value is less than the specified minimum, so M_{MAT} is set to 0.1. The third factor to be applied is based on the release temperature vs. normal boiling point, as described in Section 2.8.2.5. It is expected that the material will flash to its boiling point upon release, so this factor is set to 1.

The last multiplier accounts for an indoor release, and in Level 3 the ventilation can be taken into account, given the following inputs:

Building volume (V) = 30 ft × 50 ft × 20 ft = 30,000 ft^3

Ventilation rate (EVR) = 3 air changes per hour. It is further assumed that there is no advanced ventilation system in place that would increase the rate of ventilation upon detection of a release.

Draft direction—It is assumed that the draft is not specifically designed to draw vapors away from likely ignition sources inside the building (B_{vdd} = 1).

Per Appendix B, the indoor multiplier M_V is then

$$M_V = 1.5 \times B_{es} \times B_{vr} \times B_{vdd}$$

$$B_{es} = (V/150,000)^{-1/3} = (30,000/150,000)^{-1/3} = 1.71$$

$$B_{vr} = (EVR/2)^{-1/2} = (3/2)^{-1/2} = 0.816$$

$$M_V = 1.5 \times 1.71 \times 0.816 \times 1 = 2.09$$

Lastly,

$$PODI_{Level\ 3inside} = PODI_{S/D} \times M_{MAG} \times M_{MAT} \times M_T \times M_V$$

$$PODI_{Level\ 3inside} = 0.377 \times 1 \times 0.1 \times 1 \times 2.09 = 0.079$$

Now the contribution of the outside ignition sources will be considered.

Diesel Train

The baseline PODI is based on ignition source strength and event duration. The source strength, as listed in Table 2.2, is S = 0.4 (for the time being it will be assumed that the trains are always active). The baseline PODI is calculated as per Section 2.8.2.2:

$$PODI_{S/D} = 1 - [(1 - S^2) \times e^{-St}]$$

$$PODI_{S/D} = 1 - [(1 - 0.4^2) \times e^{-0.4 \times 3}] = 0.747$$

Most of the modifiers are the same as in the inside calculation:

$$M_{MAG\_Hole\ Diameter\ (liquid)} = 1;\ M_{MAT} = 0.1;\ M_T = 1$$

In this case, the ignition is being evaluated outdoors, and so $M_{IN/OUT}$ = 1. Then,

$$PODI_{Level\ 3train} = PODI_{S/D} \times M_{MAG} \times M_{MAT} \times M_T \times M_{IN/OUT}$$

$$PODI_{Level\ 3train} = 0.747 \times 1 \times 0.1 \times 1 \times 1 = 0.0747$$

Also, since this ignition source is present only 1% of the time, a 0.01 factor is applied to the result above, giving a $PODI_{Level\ 3train} = 0.000747$.

Motor Vehicles

The baseline PODI is based on ignition source strength and event duration. The source strength, as listed in Table 2.2, is $S = 0.3$ (for the time being it will be assumed that there is always a "live" vehicle present). The baseline PODI is calculated as per Section 2.8.2.2:

$$PODI_{S/D} = 1 - [(1 - S^2) \times e^{-St}]$$
$$PODI_{S/D} = 1 - [(1 - 0.3^2) \times e^{-0.3 \times 3}] = 0.630$$

The other modifiers are the same as for the train case:

$M_{MAG\_Hole\ Diameter\ (liquid)} = 1$; $M_{MAT} = 0.1$; $M_T = 1$; $M_{IN/OUT} = 1$.

Then,

$$PODI_{Level\ 3train} = PODI_{S/D} \times M_{MAG} \times M_{MAT} \times M_T \times M_{IN/OUT}$$
$$PODI_{Level\ 3train} = 0.630 \times 1 \times 0.1 \times 1 \times 1 = 0.0630$$

Since this ignition source is present only 5% of the time, a 0.05 factor is applied to the result above, giving a $PODI_{Level\ 3train} = 0.00315$.

Sum of All Delayed Ignition Sources

Strictly speaking, the individual delayed ignition sources are not additive, since ignition by one source precludes ignition by the others. In this case, the inside ignition dominates the analysis, and so adding the results is appropriate (ignoring significant figures for the purposes of illustration):

[Approximate (additive) calculation]

$PODI_{Level\ 3combined} \sim 0.079 + 0.000747 + 0.00315 = 0.0829$

[Exact calculation]

$PODI_{Level\ 3combined} = 0.079 + (1 - 0.079)(0.000747)$
$$+ \{1 - [0.079 + (1 - 0.079)(0.000747)]\}(0.00315) = 0.0826$$

Final Level 3 Ignition Probability Calculations

POII was estimated above to be 0.152. Since immediate ignition precludes the possibility of a delayed ignition, the probability of delayed ignition is

$$PODI_{final} = PODI_{calc}\,(1 - POII) = 0.0826\,(1 - 0.152) = 0.070$$

This PODI includes both fire and explosion outcomes.

Discussion of Results

This case suggests two possible approaches to the analysis—to model the release as a liquid or to consider only the vapors that evolve from the spill and consider the event purely from a vapor release perspective. In principle, modeling the event with respect only to the generated vapors should be more accurate, since the purely liquid model has to crudely infer the resulting vapor cloud characteristics (since this is, after all, not a consequence modeling tool). However, the algorithms developed for this book permit use of liquid inputs on the basis that an analyst may not have a sophisticated discharge/dispersion model at hand.

An extension of this example is provided in Section 4.7.4, where the benefit of having an advanced ventilation system is assessed.

4.5.2   Release of Ammonia

Ammonia is only marginally flammable but has potential to explode if released indoors. It is also toxic, with an IDLH of 300 ppm. A Level 1 analysis has been performed, which is fed into an event tree that described the frequency and outcomes of both flammable and toxic outcomes (Figure 4.9). For comparison purposes, a Level 2 analysis is also performed and fed into the same outcome/risk event tree (Figure 4.10).

In this case, using the "more conservative" ignition model in Level 1 results in nonconservative risk results, since the lower severity ignition events preclude the higher severity toxic outcomes from occurring. Thus, as suggested in Case 2 in Section 4.1.2, a "conservative" ignition model does not necessarily result in conservative overall results for a release that is both flammable and toxic.

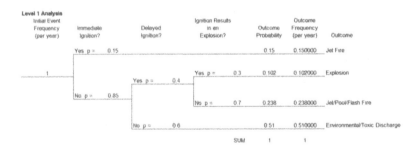

Figure 4.9. Event tree for Level 1 ammonia release.

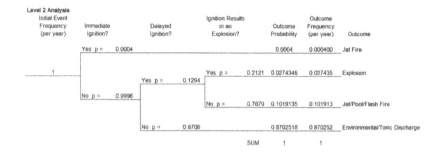

Figure 4.10. Event tree for Level 2 ammonia release.

4.6 WORKED EXAMPLES ("OUT OF SCOPE" CASES)

These "out of scope" cases represent situations that are qualitatively different in some manner from the intended scope of this book (onshore releases of flammables). The intent in presenting these examples is solely to illustrate the degree to which the results for these "out of scope" applications may be similar to or different than the applications that are in scope. *The tools in this book should* ***not*** *be expected to provide accurate results for these situations.*

4.6.1 Release of Gas from an Offshore Platform Separator

4.6.1.1 The Event Description

This event involves the release of separator gas from an offshore production deck, specifically a leak or rupture from a separator overhead line on the production deck. P = 100 barg, T = 30 °C, and composition is methane rich, but with significant quantities of heavier gases. The release can be detected and isolated within 1 minute (blowdown will be activated upon detection), but this kind of release can fill the module volume in that period of time depending on the size of the hole. The piping is 30 inches in diameter and is assumed to rupture.

The rupture/blowdown will result in a rapid depressurization of the system, and this affects the pressure input. It is also appropriate to consider, during the lower pressure period, whether it is appropriate to use the 30-inch hole size as a proper input, since the hole size is a surrogate for the release rate—which is, of course, affected by the pressure (more on that during the calculations).

The assumed pressure profile is as follows:

Time (min)	Pressure (barg)
0	100
1	50
2	25
10	10

The ignition sources are typical for offshore platforms and are assumed to be equivalent to a "high-density process area" as described in Table 2.2.

4.6.1.2   Ignition Probability Calculations

Level 2 Immediate Ignition Probability Calculations

As per Section 2.8.1.1, the "static" contribution to immediate ignition can be expressed as follows after combining the individual contributors and modifiers:

$$POII_{static} = 0.003 \times P^{1/3} \times \{MIE \times exp[0.0044(60 - T)]\}^{-0.6}$$

P is 100 barg (1450 psig) and T is 30 °C (86 °F). The release is a mixture of methane and other gases with an average MIE of 0.29 mJ. Then

$$POII_{static} = 0.003 \times (1450)^{1/3} \times \{(0.29) \times exp[0.0044(60 - 86)]\}^{-0.6}$$

$$= 0.076$$

Section 2.8.1.2 describes the contribution of autoignition. Since the temperature of this stream is well below its AIT, there is no contribution of autoignition to the POII. The predicted overall POII is therefore 0.076.

Level 2 Delayed Ignition Probability Calculations

The "high-density process" area ignition source has a value of strength S of 0.25, as per Table 2.2. The PODI is then calculated using the relationship in Section 2.8.2.2:

$$PODI_{S/D} = 1 - [(1 - S^2) \times e^{-St}]$$

$$PODI_{S/D} = 1 - [(1 - 0.25^2) \times e^{-0.25 \times 10}] = 0.923$$

The first factor to be applied to the baseline PODI is the "Magnitude of Release" multiplier. This is estimated using the diameter of the failed pipe, as described in Section 2.8.2.3. Using the hole size provided gives the following:

$$M_{MAG\_Hole\ Diameter\ (liquid)} = (Hole\ Diameter)^{0.6} = 30^{.6} = 7.70$$

However, since the hole diameter is a surrogate for release rate, one might consider artificially reducing the diameter used above to give a measure that reflects an equivalent release rate; e.g.:

Time (min)	Pressure (barg)	Equivalent Hole Size (in.)	Equivalent M_{MAG}
0	100	30	7.7
1	50	22	6.4
2	25	14	4.9
10	10	9	3.7

The user can choose the method they think is most appropriate, including calculating discrete PODIs for each time interval. However, this is all well beyond the capability of the software tool.

Note that the preceding discussion on M_{MAG} is presented for illustration purposes only; by rule M_{MAG} is limited to a value of 3.

The second multiplier to the PODI is based on the MIE of the material being released.

Per Section 2.8.2.4,

$$M_{MAT} = 0.5 - 1.7 \log(MIE), \text{ or}$$

$$M_{MAT} = 0.5 - 1.7 \log(0.29) = 1.41$$

The third factor to be applied is based on the release temperature vs. normal boiling point, as described in Section 2.8.2.5. In this case, the release is a vapor, and so the value of M_T is 1. The release is outdoors, so the indoor/outdoor multiplier described in Section 2.8.2.6 is also 1.

The combined PODI for the first ignition source is then

$$PODI_{Level\ 2} = PODI_{S/D} \times M_{MAG} \times M_{MAT} \times M_T \times M_{IN/OUT}$$

$$PODI_{Level\ 2} = 0.923 \times 3 \times 1.41 \times 1 \times 1 = 3.90$$

This value is greater than 1, suggesting that ignition is certain.

Level 2 Probability of Explosion Calculations (Example of Appendix B Algorithms)

As described in Appendix B, there are three factors that are used as multipliers onto a "base" value of 0.3 for probability of explosion given delayed ignition (POEGDI):

$$POEGDI_{Level\ 2} = 0.3 \times M_{CHEM} \times M_{MAGE} \times M_{IN/OUT}$$

These terms are calculated as follows:

M_{CHEM}—M_{CHEM} is a function of the chemical's "reactivity," or propensity to ignite explosively as measured by its fundamental burning velocity. Pure methane is a "low-reactivity" material, but this mixture has enough heavier elements that it should be treated as "medium reactivity." Therefore M_{CHEM} has a value of 1.

M_{MAGE}—M_{MAGE} is similar to M_{MAG} but is not considered as strong an influence as for PODI. Per Appendix B,

$$M_{MAGE} = (PODI\ M_{MAG})^{0.5}$$
$$M_{MAGE} = (3)^{0.5} = 1.73$$

$M_{IN/OUT}$—$M_{IN/OUT}$ is the multiplier used if a release is indoors and is calculated as for PODI. For this outdoor release, the value is 1.

Thus the overall Level 2 POEGDI is

$$POEGDI_{Level\ 2} = 0.3 \times 1 \times 1.73 \times 1 = 0.52$$

Final Level 2 Ignition Probability Calculations

POII was estimated above to be 0.076. Since immediate ignition precludes the possibility of a delayed ignition, the probability of delayed ignition is

$$PODI_{final} = PODI_{calc}\ (1 - POII) = 1(1 - 0.076) = 0.924$$

This PODI includes both fire and explosion outcomes. The fraction of delayed ignitions that result in explosions is calculated above as 0.52. Therefore the probabilities of each outcome are:

POII = 0.076

PODI resulting in an explosion = 0.924(0. 52) = 0.48

PODI resulting in a fire only = 0.924(0.48) = 0.44

Discussion of Results

The company's experience with this type of event is that it will most likely explode if it ignites at all. The calculations here appear to be somewhat consistent with that observation, although one can hardly generalize based on this one test that the methods in this book would be appropriate for all offshore situations.

4.6.2   Dust Ignition

This is a good example of a case that is clearly beyond the scope of this book. Nonetheless, it is useful to illustrate the difficulties of predicting dust explosion frequencies.

Probably one of the most difficult aspects of predicting dust explosion frequencies is determining the frequency of the initiating event—which is generally considered to be something that creates a disturbance in an otherwise stable collection of dust. Assuming that this frequency can be estimated by some means, this example will attempt to estimate the probability that the initial disturbance of dust will result in an ignition.

4.6.2.1   The Event Description

A combustible dust having a bulk density of 28 lb/ft^3 is assumed to collect to a depth of 1/8 inch across the entire surface area of a 50 ft × 50 ft × 30 ft tall building but suspended above the room's floor on various beams, equipment, etc. The volume/density of the dust results in a total mass of 730 pounds, which is more than enough to fill the room to the dust's lower explosive limit of 30 g/m^3. An initiating event occurs that greatly disturbs the dust but does not immediately ignite it.

A Level 2 analysis is being performed to estimate the probability that the dust will ultimately ignite/explode, but the analysis will be performed for both high-particle-size dusts (with assumed MIE of 1,000 mJ) and small-particle-size dusts (assumed MIE 1 mJ). The time frame over which the dust cloud would persist in the open-room environment is assumed to be 2 minutes for the small-particle-size dust and 30 seconds for the large-particle-size dust.

4.6.2.2   Potential Ignition Sources

The following ignition source is present:

Process Area—The area in which this event occurs can be described as a "low-density process area" in the sense that there is not an abundance of ignition sources present.

4.6.2.3   Ignition Probability Calculations

Level 2 Immediate Ignition Probability Calculations

As per Section 2.8.1.1, the "static" contribution to immediate ignition can be expressed as follows after combining the applicable individual contributors and modifier:

$$POII_{static} = 0.003 \times P^{1/3} \times MIE^{-0.6}$$

In this case, the pressure in the room is 0 psig; therefore $POII_{static} = 0$. Autoignition is not an issue here; therefore the overall predicted POII is 0, as one would expect (the secondary ignition being modeled here would probably occur in the time period proximate to the initial "disturbance" event; however, ignition occurs from mechanisms other than those treated by the immediate ignition algorithms and so is evaluated as a delayed ignition).

Level 2 Delayed Ignition Probability Calculations

There is a single area ignition source to consider. The source strength, as listed in Table 2.2, is $S = 0.1$. The baseline PODI is calculated as per Section 2.8.2.2, as applied to our two dust scenarios:

$$PODI_{S/D} = 1 - [(1 - S^2) \times e^{-St}]$$

$$PODI_{S/Dfinedust} = 1 - [(1 - 0.1^2) \times e^{-0.1 \times 2}] = 0.189$$

$$PODI_{S/Dcoarsedust} = 1 - [(1 - 0.1^2) \times e^{-0.1 \times 0.5}] = 0.058$$

The first factor to be applied to the baseline PODI is the "Magnitude of Release" multiplier. This can be estimated using either the diameter of the failed pipe or the amount released, as described in Section 2.6.2.3. For the purposes of this example, it will be assumed that the dust is fine enough to be treated as a vapor. Then

$$M_{MAG\_Amount\ Released(vapor)} = (Amount\ Released/1,000)^{0.5} = (730/1,000)^{0.5} = 0.854$$

The second multiplier to the PODI is based on the MIE of the material being released. Per Section 2.8.2.4:

$$M_{MAT} = 0.5 - 1.7\ log(MIE),\ or$$

$$M_{MATfinedust} = 0.5 - 1.7\ log(1) = 0.5$$

$$M_{MATcoarsedust} = 0.5 - 1.7\ log(1000) = -4.6$$

Since the value for the coarse dust is negative, it is reset to the minimum value of 0.1, as per Section 2.8.2.4.

The third factor to be applied is based on the release temperature vs. normal boiling point, as described in Section 2.8.2.5, but this does not apply to vapor (or dust) releases. Also, the release is indoors, so the indoor/outdoor multiplier described in Section 2.8.2.6 is 1.5.

The combined PODI for the first ignition source is then

$$PODI_{Level\ 2} = PODI_{S/D} \times M_{MAG} \times M_{MAT} \times M_T \times M_{IN/OUT}$$

$$PODI_{Level\ 2,finedust} = 0.189 \times 0.854 \times 0.5 \times 1 \times 1.5 = 0.121$$

$$PODI_{Level\ 2,coarsedust} = 0.058 \times 0.854 \times 0.1 \times 1 \times 1.5 = 0.0074$$

Level 2 Probability of Explosion Calculations (Test of Appendix B)

As described in Appendix B, there are three factors that are used as multipliers onto a "base" value of 0.3 for probability of explosion given delayed ignition (POEGDI):

$$POEGDI_{Level\ 2} = 0.3 \times M_{CHEM} \times M_{MAGE} \times M_{IN/OUT}$$

These terms are calculated as follows:

M_{CHEM}—M_{CHEM} is a function of the chemical's "reactivity," or propensity to ignite explosively as measured by its fundamental burning velocity. In practice, this is a function of a number of variables including the chemical composition of the dust ("Kst" could be a reasonable measure) and its particle size. For the purposes of this exercise, the fine dust is considered a "medium-reactivity" material, and the coarse dust "low reactivity." Therefore, M_{CHEM} has values of 1 and 0.5, respectively.

M_{MAGE}—M_{MAGE} is similar to M_{MAG} but is not considered as strong an influence as for PODI. Per Appendix B,

$$M_{MAGE} = (PODI\ M_{MAG})^{0.5}$$

$$M_{MAGE} = (0.854)^{0.5} = 0.924$$

$M_{IN/OUT}$—$M_{IN/OUT}$ is the multiplier used if a release is indoors and is calculated as for PODI. For this indoor release, the value is 1.5.

Thus the overall Level 2 POEGDI is

$$POEGDI_{Level\ 2,finedust} = 0.3 \times 1 \times 0.924 \times 1.5 = 0.423$$

$$POEGDI_{Level\ 2,coarsedust} = 0.3 \times 0.5 \times 0.924 \times 1.5 = 0.212$$

Final Level 2 Ignition Probability Calculations

POII was estimated above to be 0. The PODI includes both fire and explosion outcomes. The fraction of delayed ignitions that result in explosions is calculated above as 0.423 and 0.212 for the fine and coarse dusts, respectively. Therefore the probabilities of each outcome are:

$POII_{fine\ dust} = 0$

PODI resulting in an explosion$_{fine\ dust}$ = 0.121(0.423) = 0.051

PODI resulting in a fire only$_{fine\ dust}$ = 0.121(0.577) = 0.070

$POII_{coarse\ dust} = 0$

PODI resulting in an explosion$_{coarse\ dust}$ = 0.0074(0.212) = 0.0016

PODI resulting in a fire only$_{coarse\ dust}$ = 0.0074(0.788) = 0.0058

Discussion of Results

One can debate the validity of these results. In any case, as noted in the opening to this example, probably the more problematic issue is estimating the frequency of the initial "disturbance" event that caused the accumulated dust to be dislodged. This is an area that deserves more evaluation but is beyond the scope of this book to address.

4.7 WORKED EXAMPLES OF THE BENEFITS OF PLANT MODIFICATIONS AND DESIGN CHANGES

The following examples illustrate the use of the algorithms in this book to calculate the benefit of incorporating improved equipment or systems for flammables release management or in selecting among design alternatives.

4.7.1 Ignition by Hot Surfaces

4.7.1.1 Event Description

A risk sensitivity analysis is being conducted to determine an appropriate driver for the refrigeration (propane) gas compressor for a gas processing plant. Three driver options are being considered—motor, steam, and combustion turbine, each with different surface temperatures as shown below.

Item	Surface Temperature (°F)
Motor	150
Steam turbine	750
Combustion turbine	1500

The combustion turbine is shut down upon the detection of leaked hydrocarbon; however, because of its long coast-down time, a flammable mixture may be sucked into the intake air manifold.

A worst-case, credible equivalent diameter hole size of 2 inches is used as a basis in the analysis. The compressor discharge process conditions are 315 psig at a temperature of 180 °F. A process simulator shows that an equivalent 2-inch-hole leak at this location causes a flow rate of 25 lb/s of propane. A dispersion analysis using a wind speed of 5 m/s and a class D stability shows that as the cloud drifts downwind, the vapor cloud quickly rises to ambient air conditions of 50 °F. The downwind distance to the LFL (20,000 ppm) is 100 feet and so is capable of fully engulfing the compressor driver, which is only 30 feet from the release location.

For the purposes of this analysis, the other equipment items in the plant are assumed to provide relatively negligible ignition sources. A Level 2 analysis is to be performed.

4.7.1.2   Ignition Probability Calculations

Level 2 Immediate Ignition Probability Calculations

As per Section 2.8.1.1, the "static" contribution to immediate ignition can be expressed as follows after combining the individual contributors and modifiers:

$$POII_{static} = 0.003 \times P^{1/3} \times \{MIE \times exp[0.0044(60 - T)]\}^{-0.6}$$

P is 315 psig and T is 180 °F. The MIE for propane is 0.25 mJ. Therefore,

$$POII_{static} = 0.003 \times (315)^{1/3} \times \{(0.25) \times exp[0.0044(60 - 180)]\}^{-0.6}$$

$$= 0.064$$

Section 2.8.1.2 describes the contribution of autoignition. Since the temperature of this stream is well below its AIT, there is no contribution of autoignition to the POII. The predicted overall POII is therefore 0.064.

Level 2 Delayed Ignition Probability Calculations

Section 2.3.1 describes the treatment of hot surfaces. The ignition source strength S is estimated as

$$S = 0.5 + 0.0025 [T - AIT - 100(CS)]$$

where CS is the cloud speed and T is the temperature of the hot surface. Although there is likely still some jet behavior in the cloud at the distance of the ignition source, the calculation will be treated conservatively and the cloud speed will be assumed to be equal to the wind speed of 5 m/s. For the three design options, S can be estimated as

$$S_{motor} = 0.5 + 0.0025 [150 - 842 - 100(5)] = -2.5$$

$$S_{steam\ turbine} = 0.5 + 0.0025 [750 - 842 - 100(5)] = -1.0$$

$$S_{combustion\ turbine} = 0.5 + 0.0025 [1500 - 842 - 100(5)] = 0.89$$

The results for the motor and the steam turbine indicate that hot-surface ignition should not be significant in either case. However, the combustion turbine is hot enough to be of concern.

It is assumed that the ignition conditions are active for 3 minutes. The baseline PODI for the combustion turbine is then calculated as per Section 2.6.2.2:

$$PODI_{S/D} = 1 - [(1 - S^2) \times e^{-St}]$$

$$PODI_{S/D} = 1 - [(1 - 0.89^2) \times e^{-0.89 \times 3}] = 0.986$$

The first factor to be applied to the baseline PODI is the "Magnitude of Release" multiplier. This is estimated using the diameter of the failed pipe, as described in Section 2.8.2.3:

$$M_{MAG\_Hole\ Diameter\ (liquid)} = (\text{Hole Diameter}) = 2$$

The second multiplier to the PODI is based on the MIE of the material being released. Per Section 2.8.2.4,

$$M_{MAT} = 0.5 - 1.7\ \log(\text{MIE}),\ \text{or}$$

$$M_{MAT} = 0.5 - 1.7\ \log(0.25) = 1.52$$

The third factor to be applied is based on the release temperature vs. normal boiling point, as described in Section 2.8.2.5. In this case, the release is a gas, and so a multiplier of 1 is used. Similarly, the release is outdoors, so the indoor/outdoor multiplier described in Section 2.8.2.6 is 1.

The combined PODI for the first ignition source is then

$$PODI_{Level\ 2} = PODI_{S/D} \times M_{MAG} \times M_{MAT} \times M_T \times M_{IN/OUT}$$

$$PODI_{Level\ 2} = 0.986 \times 2 \times 1.52 \times 1 \times 1 = 3.0$$

Since the result is greater than 1, it should be assumed that the release will always ignite—*when oriented in the direction of the hot surface.*

Final Level 2 Ignition Probability Calculations

POII was estimated above to be 0.064. Since immediate ignition precludes the possibility of a delayed ignition, the probability of delayed ignition is

$$PODI_{final} = PODI_{calc}\ (1 - POII) = 1\ (1 - 0.064) = 0.936$$

This PODI includes both fire and explosion outcomes.

Alternative Case

The site has a methane gas compressor with similar exposures. This compressor operates at 600 psig and 150 °F. Using the same approach as above gives results for delayed ignition probability that are dramatically different than in the original case, primarily because methane has a substantially higher AIT than propane).

4.7.2   Release Prevention

Of course, the "best" release is the one that never occurs. API's risk-based inspection (RBI) protocols (API, 2000) and other RBI and non-RBI methods exist that can be used to reduce the predicted failure rate of piping and vessels. However, it must be noted that RBI addresses "inspectable" risks and not all risks. Therefore, someone attempting to use RBI as a basis for reducing release event frequencies should be very well versed in the background of the methods.

Note that the use of such methods affects, not the predicted *probability* of ignition (and so is out of the scope of this book), but rather the *frequency* of the initial release. However, these methods can be incorporated along with ignition probability reductions as part of an overall flammable risk reduction strategy.

4.7.3   Duration of Exposure

One method being used increasingly to manage flammable hazards is isolation of the release. The isolation may be performed automatically or manually following detection of a flammable cloud and results in the remaining process fluid being contained in the process. Alternatively, the release may be diverted to a closed sump area so that it does not contact an ignition source. The former case is evaluated next.

4.7.3.1   Event Description

Isobutane is loaded from railcars into a storage bullet at a chemical plant site. The transfer takes place using a 3-inch loading arm; there is concern that the loading arm could fail during an off-load for one of several reasons that have been identified in a FMEA study. Although the event is very unlikely, the consequences are potentially catastrophic given the people and processes in the vicinity. At present, the release could continue for several minutes before it was detected and responded to completely. The site wants to evaluate whether measures such as an excess flow valve or rapid detection/shutdown/isolation would be of significant benefit. For the purposes of the analysis, it is assumed that isolation could be completed in 1 minute using the advanced isolation methods.

A worst-case, credible equivalent diameter hole size of 3 inches is used as a basis in the analysis. The isobutane transfer occurs at 100 psig and ambient temperature of 70 °F. A discharge model estimates that this event would result in an initial release rate of 200 lb/s of isobutane; it is assumed that this initial release rate can be sustained for the duration of the event. The gas cloud enters a surrounding plant area that can be described as a "high-density" process area.

A comparison is to be made at Level 2, assuming both 1-minute and 10-minute durations.

4.7.3.2   Ignition Probability Calculations

Level 2 Immediate Ignition Probability Calculations

As per Section 2.8.1.1, the "static" contribution to immediate ignition can be expressed as follows after combining the individual contributors and modifiers:

$$POII_{static} = 0.003P^{1/3} \times \{MIE_{reported} \times (10,000/P_{liquid})^{0.25} \exp[0.0044(60 - T)]\}^{-0.6}$$

P is 100 psig and T is 70 °F. The MIE for isobutane is 0.26 mJ. Therefore,

$$POII_{static} = 0.003 \times (100)^{1/3} \times \{(0.26) \times (10,000/100)^{0.25} \exp[0.0044(60 - 70)]\}^{-0.6}$$

$$= 0.016$$

Section 2.8.1.2 describes the contribution of autoignition.   Since the temperature of this stream is well below its AIT, there is no contribution of autoignition to the POII. The predicted overall POII is therefore 0.016.

Level 2 Delayed Ignition Probability Calculations

A single area ignition source is defined; as per Table 2.2, the ignition source strength S is estimated as 0.25. The baseline PODI for the combustion turbine is then calculated as per Section 2.6.2.2:

$$PODI_{S/D} = 1 - [(1 - S^2) \times e^{-St}]$$

$$PODI_{S/D, \, 1\text{-minute}} = 1 - [(1 - 0.25^2) \times e^{-0.25 \times 1}] = 0.270$$

$$PODI_{S/D, \, 10\text{-minute}} = 1 - [(1 - 0.25^2) \times e^{-0.25 \times 10}] = 0.923$$

The first factor to be applied to the baseline PODI is the "Magnitude of Release" multiplier. This is estimated using the diameter of the failed pipe, as described in Section 2.8.2.3:

$$M_{MAG\_Hole \, Diameter \, (liquid)} = (\text{Hole Diameter}) = 3$$

The second multiplier to the PODI is based on the MIE of the material being released. As per Section 2.8.2.4:

$$M_{MAT} = 0.5 - 1.7 \log(MIE), \text{ or}$$

$$M_{MAT} = 0.5 - 1.7 \log(0.26) = 1.49$$

The third factor to be applied is based on the release temperature vs. normal boiling point, as described in Section 2.8.2.5. In this case, the boiling point is well below the release temperature and so a multiplier of 1 is used. Similarly, the release is outdoors, so the indoor/outdoor multiplier described in Section 2.8.2.6 is 1.

The combined PODI for the first ignition source is then

$$PODI_{Level\ 2} = PODI_{S/D} \times M_{MAG} \times M_{MAT} \times M_T \times M_{IN/OUT}$$

$$PODI_{Level\ 2,\ 1-minute} = 0.270 \times 3 \times 1.49 \times 1 \times 1 = 1.20$$

$$PODI_{Level\ 2,\ 10-minute} = 0.923 \times 3 \times 1.49 \times 1 \times 1 = 4.12$$

Discussion of Results

Since the results are both greater than 1, it should be assumed that the release will always ignite. At this point, one might conclude that since ignition occurs at either duration, there is no benefit to providing the advanced isolation system. However, the following perspectives should also be considered:

- While there is no demonstrable benefit to providing advanced isolation for a loading arm failure case, there are lesser events for which there would be a tangible benefit, as the M_{MAG} term in the equation above is reduced.

- If the advanced isolation does not provide value, this suggests that an alternative approach should be taken; for example:

  - Focus on preventing the loading arm failure (e.g., through inspections/ replacement) rather than mitigating its effects. The original FMECA should be revisited to determine which failure causes can be reduced or eliminated.

  - Relocating the unloading station so that in the event of a failure it does not discharge into a "high-density" area.

4.7.4   Benefit of Improved Ventilation of Indoor Releases— Continuation of "Indoor Acid Spill" Example

Section 4.5.1 illustrated a case where an indoor release could contact ignition sources both inside and outside a building. In that example, however, the ignition probability was dominated by the indoor ignition source for the situation where the normal ventilation rate was 3 air changes per hour (ACH) in winter.

It may not be practical to have a higher air change rate on a continuous basis in the winter in this facility due to the difficulty in maintaining a sufficiently warm air temperature to provide an acceptable working environment (for personnel and possibly some equipment/chemicals). However, it has been determined that a temporary increase in ventilation would be preferable to accumulation of flammable gases if it can be demonstrated that this ventilation would significantly lower the probability of ignition.

A heating, ventilation, and air-conditioning (HVAC) company has determined that it would be practical to provide a ventilation rate of 15 ACH upon detection of a release. Detectors capable of detecting 10% LFL condition will be spaced every 15 feet in the room, and the ventilation can be arranged so that the air is drawn away mainly from the primary ignition sources in the room.

The original calculation for the ventilation multiplier was the following:

$$M_V = 1.5 \times B_{es} \times B_{vr} \times B_{vdd}$$

$$B_{es} = (V/150,000)^{-1/3} = (30,000/150,000)^{-1/3} = 1.71$$

$$B_{vr} = (EVR/2)^{-1/2} = (3/2)^{-1/2} = 0.816$$

$$M_V = 1.5 \times 1.71 \times 0.816 \times 1 = 2.09$$

$$PODI_{Level\ 3inside,\ current\ HVAC} = PODI_{S/D} \times M_{MAG} \times M_{MAT} \times M_T \times M_V$$

$$PODI_{Level\ 3inside} = 0.377 \times 1 \times 0.1 \times 1 \times 2.09 = 0.079$$

For the proposed ventilation system, the calculation for B_{es} remains the same, but B_{vr} and B_{vdd} get revised as follows, as per Appendix B:

$$EVR = 3 \times IVR = 9\ ACH\ and\ B_{vr} = (EVR/2)^{-1/2} = (9/2)^{-1/2} = 0.471$$

$$B_{vdd} = 0.5\ (draft\ away\ from\ likely\ ignition\ sources)$$

Then,

$$M_V = 1.5 \times 1.71 \times 0.471 \times 0.5 = 0.605$$

$$PODI_{Level\ 3inside} = PODI_{S/D} \times M_{MAG} \times M_{MAT} \times M_T \times M_V$$

$$PODI_{Level\ 3inside,\ revised\ HVAC} = 0.377 \times 1 \times 0.1 \times 1 \times 0.605 = 0.023$$

The outside ignition sources still exist. Revising the earlier calculation for the overall PODI results in

[approximate (additive) calculation]

$PODI_{Level\ 3combined} \sim 0.023 + 0.000747 + 0.00315 = 0.0267$, versus the original value of 0.0829.

Discussion of Results

This example shows that the additional expenditure for the advanced ventilation system would be expected to reduce delayed ignition probabilities (in winter) by over 68%, assuming the gas detectors and emergency ventilation activate in time. This may or may not be sufficient justification for management to authorize the funds but does at least provide an objective basis for making a decision.

5 SOFTWARE ILLUSTRATION

5.1 EXPLANATION AND INSTRUCTIONS FOR SOFTWARE TOOL

The software tool provided with this book is reasonably intuitive for users familiar with Microsoft Excel®, although the primary interface is Visual Basic®. Following are some screen prints to illustrate the step-by-step process of starting the software, populating the input fields, and reading the outputs.

5.2 OPENING THE SOFTWARE TOOL

After starting the software, the first screen the user encounters is the following:

The user must choose to "Enable Content" using the button in the upper left, at which point a disclaimer appears. Click "OK" and the starting screen appears:

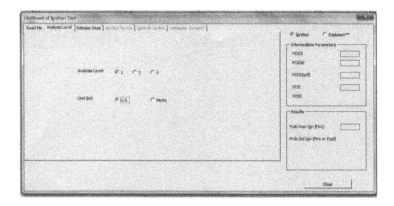

At this point the user is allowed to select the level of analysis desired (1 being basic, 2 intermediate, and 3 advanced) as well as the preferred units of measure (U.S. or metric). Next the user starts the information input process.

5.3 GENERAL INPUTS AND OUTPUTS

After selecting the desired level of analysis, the user next clicks on the "Release Data" tab:

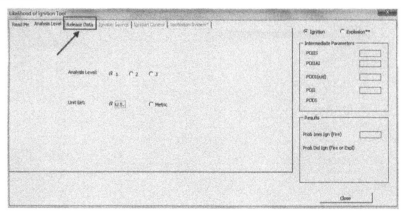

This opens the following window for the chemical pick list (note that the fields that appear in this window will depend on the level of analysis selected; a Level 1 set of inputs is illustrated here):

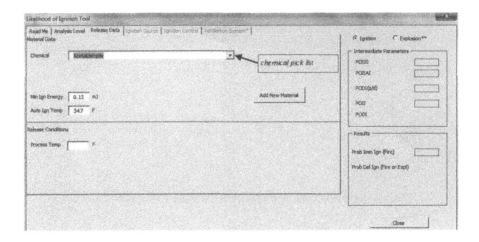

The first item from the chemical pick list is shown with its relevant physical properties. To select another chemical from the list, click the pulldown arrow on the right side of the "Chemical" field and the complete menu of chemical selections will appear in a scrolling list:

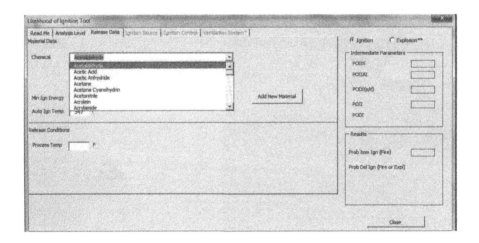

If the chemical of interest is not in the pick list, the user may manually enter a new material using the "Add New Material" button to the right of the pick list.

Users can add the new chemical properties from this screen:

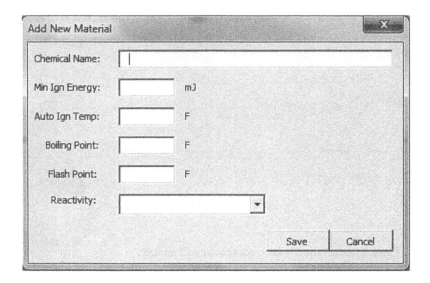

The user can now input the process conditions necessary to support the level of analysis that was originally selected. Note that the analysis level can be changed at any time, from the simplest (Level 1) to the most complex (Level 3).

5.4 LEVEL 1 INPUTS

The Level 1 inputs for the "Release Data" tab are very basic, commensurate with its expected use in a HAZOP, LOPA, or similar analyses where only low-resolution values are needed and more detailed information is not available. In fact, aside from the chemical properties (which in most cases will be pre-entered, the only additional information needed is the process temperature):

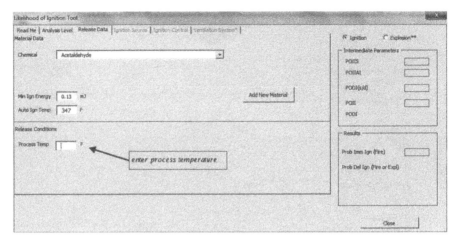

Typical Level 1 "Material Data" Screen

Once a temperature is entered, results will appear:

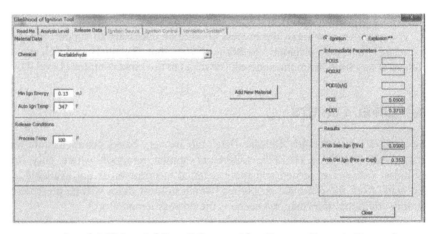

Level 1 "Material Data" Screen After Process Temp is Entered

The probability of immediate ignition (POII) and probability of delayed ignition given that immediate ignition did not occur (PODI) are shown in the "Intermediate Parameters" output box. The "Results" box shows the results as modified to reflect the fact that a delayed ignition can only occur if an immediate ignition has not. Other output fields are not visible, since they pertain to higher levels of analysis.

5.5 LEVEL 2 ANALYSES

A Level 2 analysis is similar to Level 1, but with more additional input information required. The input screens are similar to those in Level 1, but in some cases the entry of one field will cause a pop-up to appear requesting additional information. Following is an example of a Level 2 input screen:

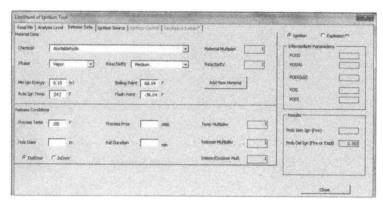

Typical Level 2 "Release Data" Screen

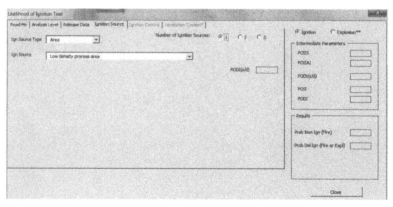

Typical Level 2 "Ignition Source" Screen

Note that depending on the type or number of ignition sources selected, the screen contents may expand as shown below:

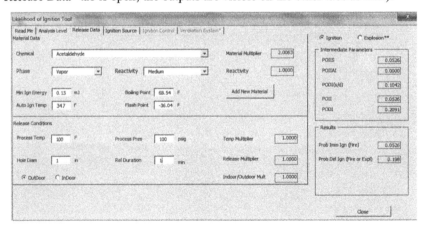

Expanded Level 2 "Ignition Source" Screen

Following is an example of a typical Level 2 output screen (in this case the "Release Data" tab is open; the outputs are visible on the other tabs as well):

Typical Level 2 Output Screen

The outputs have the same meanings as they do at Level 1; the only difference is that additional intermediate calculation values are provided in Levels 2 and 3 in the box on the right side of the screen. Refer to Chapter 2 of this book for details on the meaning of the intermediate calculations.

5.6   LEVEL 3 ANALYSES

A Level 3 analysis uses the same approach as Level 2, but with additional input options. Choosing a Level 3 analysis activates a new "Ignition Control" tab in which users enter information about systems they have in place to prevent or control ignitions. Also available on Level 3 is a more detailed analysis of indoor ignition events.

Clicking the "Indoor" button on the "Release Data" tab activates the "Ventilation System" tab in which that refinement can be performed. When selected, this tab displays a disclaimer indicating its relatively speculative nature.

5.7   EXPLOSION PROBABILITY

In a Level 2 or Level 3 analysis, explosion probability can be predicted by clicking the "Explosion" button in the upper right of the screen:

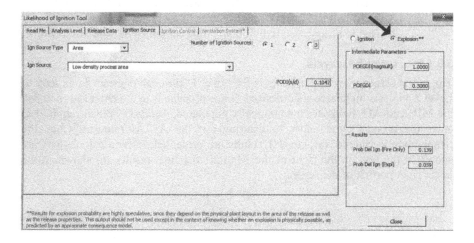

Note the disclaimer at the bottom of this screen. Do *not* rely solely on the algorithms in this tool to evaluate explosion probability.

5.8   ILLUSTRATIONS OF SOFTWARE USE

Following are some screenshots using examples from Chapter 4.

5.8.1   Vapor Cloud Explosion Hazard Assessment of a Storage Site (Example from Section 4.2.1)

This is an example of a Level 2 analysis; however, Level 1 results will also be presented. The details of the example are not repeated here, except for the following inputs to the analysis:

- Material being released—liquid propane

- Release conditions—116 psig and 68 °F

- Size of release—4-inch hole

- Event duration—4 minutes

- Ignition sources—0.5 car "active" on average, control room that is considered equivalent to "office space," substation that is considered equivalent to 20 feet of high-power line. These ignition sources will need to be assessed individually, then combined outside the software.

5.8.1.1   Level 1 Analysis

The initial screen view that appears is for Level 1, although one can go directly to Level 2 analysis input screens if desired. Since propane is in the chemical pick list, the MIE and AIT fields are automatically populated. The user is then required to enter the process temperature to determine if the AIT is relevant. Once the temperature is entered, the Level 1 results are presented. Interim calculations are shown in the box on the right of the window; the final results are shown in the lower right portion of the screen:

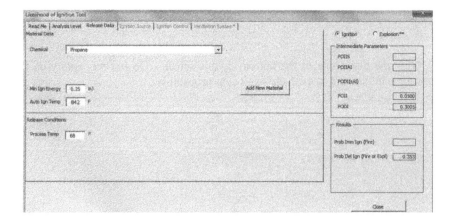

5.8.1.2   Level 2 Analysis

Progressing on to Level 2, a number of additional inputs are presented. Many fields are prepopulated since propane is in the chemical pick list. The software assumes the release is outdoors unless "Indoor" is selected instead.

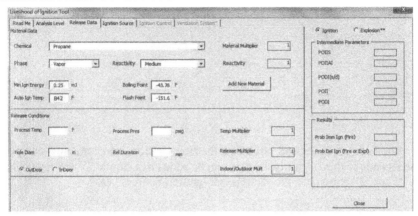

Users should carefully check the entry selections for each input field to make sure the data is correct. For example, the default Phase value is for "Vapor" and would have to be changed to liquid if that is the desired input:

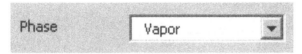

The necessary process conditions can now be entered. Note that the results are invalid at this point since an ignition source has not yet been entered:

The first ignition source to be assessed is a car, which is considered a "point source." To enter this input, the user must open the "Ignition Source" tab and enter first the type of ignition source (point source) and then the point-source type (motor vehicle) in the input boxes:

The Level 2 inputs and results for the first ignition source are now complete. The second (or third) ignition sources can be added in a similar manner by clicking the "2" or "3" buttons on the "Number of Ignition Sources" entry:

The results now appear:

$$POII = 0.0176$$

$$PODI = 0.973$$

Note that explosion results are not provided since the "Explosion" button in the upper right of the screen has not been activated.

5.8.2   Open Field Release of Propane (Example from Section 4.2.2)

This example compares the outputs from Level 1, 2, and 3 analyses. The details of the example are not repeated here, except for the following inputs to the analysis:

- Material being released—liquid propane

- Release conditions—116 psig and 68 °F

- Amount released—3,500 lb (4-inch hole)

- Event duration—3 minutes

- Ignition source—This is a tank farm area best characterized as a "Remote outdoor storage area."

5.8.2.1   Level 1 Analysis

The screen view for Level 1 is straightforward. Since propane is in the chemical pick list, the MIE and AIT fields are automatically populated. The software assumes the release is outdoors unless "Indoor" is selected instead:

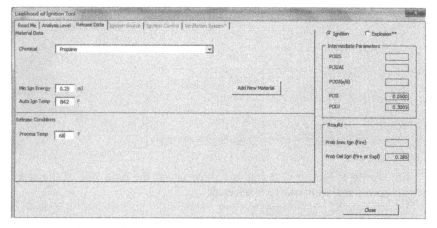

The Level 1 results also appear in the lower right corner, once the process temperature has been added in the "Release Conditions" tab.

5.8.2.2   Level 2 Analysis

Selecting a Level 2 Analysis results in expanded input screens, as seen in the previous example. These input screens contain a number of entries that do not apply to this particular analysis. Some of these (e.g., "Reactivity") are system defaults. Obviously, each input needs to be scrutinized and changed as necessary.

For this example, inputs are as shown below:

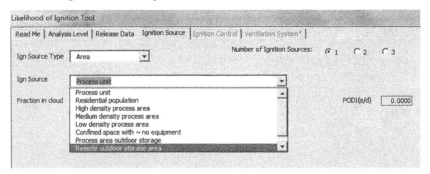

Note that this ignition source type is an "Area" source. When this source type is selected, the menu for "Ign Source" changes to allow selection of the "Remote outdoor storage area" as an input:

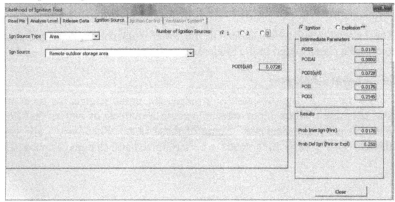

The results of the Level 2 analysis show up on the right side of the screen once the specific ignition source is selected:

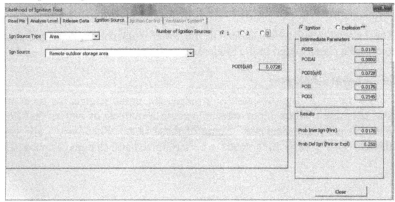

5.8.2.3   Level 3 Analysis

Choosing a Level 3 analysis results in the most expanded input screens. Note that at Level 3 the user can input a revised value for the temperature, the reasons for which are described in Sections 2.7.1 and 4.2.2.4 (in this example, the temperature will be changed to −44 °F).

In addition, Level 3 allows for an additional input for the degree of ignition source control and provides the option to manually enter the effectiveness of any special ignition or explosion control measures that might be in place, as shown in the close-up view of the bottom input fields added at Level 3:

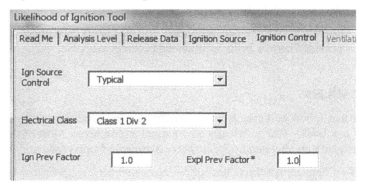

Note that default values apply. If there are no special circumstances present (as in this case), the calculations proceed on their own.

The experimental explosion ignition algorithm can also be utilized at Level 3 by clicking the "Explosion" button in the upper right:

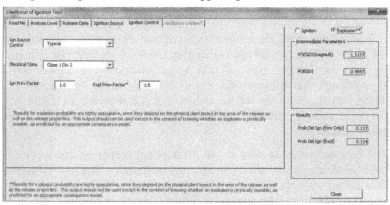

As before, the ignition probability results appear in the lower right of the screen.

APPENDIX A.
CHEMICAL PROPERTY DATA

INTRODUCTION

Various chemical property data are required to utilize the methods in this book. These data are incorporated in the software. Not all data is required for all levels of the analysis, and some of the information is provided for general interest or potential future developments.

The sources of these data are varied, and the user is encouraged to replace the values provided here as better information becomes available.

DATA TABLES

The data that follow in Table A.1 are described for the conditions that govern the scope of this book—that is, releases into normal atmospheres. Most of the entries are self-explanatory, but some specific abbreviations and terms used are:

CAS—Chemical Abstracts Service reference number

LFL—Lower flammability limit in air

Flame Speed Class—A measure of the propensity of a material to ignite explosively, as determined by its burning velocity

NFPA F—The flammability rating for the chemical as developed by the National Fire Protection Association

It is noted that there are sometimes significant differences in reported values for MIE, AIT, and flash point due to differing test methods from one experimenter to the next or in the type of test (e.g., open-cup vs. closed-cup flash point). Although the software attempts to minimize these inconsistencies, they will always exist, and some of the values that are provided in Table A.1 and utilized in the software can be expected to change over time as better information becomes available.

The data set provided here is just a sampling; properties for approximately 200 additional chemicals are built into the software. Not all properties are available for all chemicals, however; in some cases the user may need to develop missing data from alternative sources or through laboratory testing.

Table A.1. Ignition and other properties of common chemicals

Chemical	CAS Number	Mol Wt	Boiling Point (°C)	Flash Point (°C)	Auto-ignition Temp (°C)	Min. Ignition Energy (mJ)	Flame Speed Class	Flame Speed (cm/s)	Easily Ignited	LFL (vol%)	NFPA F
Acetaldehyde	75-07-0	44.05	20.3	-37.8	175	0.13	Medium		Normal	4	
Acetone	67-64-1	58.1	56.1	-17.8	465	0.19	Medium	54	Medium	2.6	3
Acrolein	107-02-8	56.1	52.5	-26	235	0.13	Medium	66	Medium	2.8	3
Acrylonitrile	107-13-1	53.1	77.2	-5	481	0.16	Medium	50	Medium	3	3
Allyl Chloride	107-05-1	76.53	45	-32	485	0.77	Medium		Medium	2.9	3
Ammonia	7664-41-7	17.03	-33.4	-65	650	680	Low		Low	15	1
Benzene	71-43-2	78.1	80.1	-11	498	0.2	Medium	48	Medium	1.2	3
Butadiene (1,3)	106-99-0	54.1	-4.4	-76.2	420	0.13	Medium	68	Medium	2	4
Butane	106-97-8	58.1	-0.6	-72	370	0.25	Medium	45	Medium	1.6	4
Carbon Disulfide	75-15-0	76.14	46.2	-30	90	0.009	High	58	High	1.3	4

Table A.1. Ignition and other properties of common chemicals, continued

Chemical	CAS Number	Mol Wt	Boiling Point (°C)	Flash Point (°C)	Auto-ignition Temp (°C)	Min. Ignition Energy (mJ)	Flame Speed Class	Flame Speed (cm/s)	Easily Ignited	LFL (vol%)	NFPA F
Di-isobutylene	107-39-1	112.2	101.4	-5	391	0.23	Medium		Medium	0.8	3
Ethane	74-84-0	30.07	-88.7	-130.2	472	0.23	Medium	47	Medium	3	4
Ethanol	64-17-5	46.07	78	13	365	0.23	Medium		Medium	3.3	3
Ethyl Acetate	141-78-6	88.11	77.1	-4	426	0.23	Medium	38	Medium	2	3
Ethyl Acrylate	140-88-5	100.1	99.7	10	372	0.18	Medium		Medium	1.4	3
Ethylene	74-85-1	28.05	-103.8	-140	450	0.084	High	80	High	2.7	4
Ethylene Oxide	75-21-8	44.05	10.5	-50	429	0.065	High	108	High	3	4
Heptane	142-82-5	100.2	98.4	-4	213	0.24	Medium	46	Medium	1.1	3
Hydrogen	1333-74-0	2.02	-252.8	-259	400	0.016	High	312	High	4	4
Hydrogen Sulfide	7783-06-4	34.08	-60.3		260	0.068	Medium		Medium	4	4

Table A.1. Ignition and other properties of common chemicals, continued

Chemical	CAS Number	Mol Wt	Boiling Point (°C)	Flash Point (°C)	Auto-ignition Temp (°C)	Min. Ignition Energy (mJ)	Flame Speed Class	Flame Speed (cm/s)	Easily Ignited	LFL (vol%)	NFPA F
Methane	74-82-8	16.04	-161.5	-187.2	537	0.21	Low	40	Low	5	4
Methanol	67-56-1	32.04	64.5	11	385	0.14	Medium	56	Medium	6.7	3
Methyl Acrylate	96-33-3	86.09	80.7	-3	468	0.18	Medium		Medium	2.8	3
Methyl Ethyl Ketone	78-93-3	72.11	79.6	-9	404	0.21	Medium		Medium	1.4	3
Methylene Chloride	75-09-2	84.93	40		556	>1000	Medium		Medium	14.5	1
Propane	74-98-6	44.1	-42.1	-102	450	0.25	Medium	46	Medium	2.1	4
Propionaldehyde	123-38-6	58.08	47.9	-30	207	0.18	Medium	58	Medium	2.9	3
Propylene	115-07-1	42.1	-47.7	-108.2	455	0.18	Medium	52	Medium	2	4
Propylene Oxide	75-56-9	58.08	34.2	-37	449	0.13	High	82	High	2.3	4
Styrene	100-42-5	104.15	145.2	31	470	0.18	Medium		Medium	0.88	3

Table A.1. Ignition and other properties of common chemicals, continued

Chemical	CAS Number	Mol Wt	Boiling Point (°C)	Flash Point (°C)	Auto-ignition Temp (°C)	Min. Ignition Energy (mJ)	Flame Speed Class	Flame Speed (cm/s)	Easily Ignited	LFL (vol%)	NFPA F
Tetrahydrofuran	109-99-9	72.1	66	-14	321	0.19	Medium		Medium	2	3
Toluene	108-88-3	92.14	110.6	4	480	0.24	Medium	41	Medium	1.1	3
Triethylamine	121-44-8	101.19	88.4	-7	249	0.22	Medium		Medium	1.2	3
Trimethylamine	75-50-3	59.1	2.8	-71	190	<0.3	Medium		Medium	2	4
Vinyl Acetate	108-05-4	86.1	72.8	-8	402	0.16	Medium		Medium	2.6	3
Xylene, m-	108-38-3	106.17	139.1	29	465	0.2	Medium		Medium	1.1	3

ALTERNATIVES FOR MIE DATA NOT PROVIDED

MIE values for additional chemicals are available from a variety of sources including NFPA 77. MIEs can also be estimated for various classes of organic compounds using the methods in Britton (2002).

APPENDIX B.
OTHER MODELS FOR CONSIDERATION

INTRODUCTION

There are models which may be useful but are not included in the main text of this book. These models are not included in the "primary" probability of ignition models either because of a technical limitation or because the model has had only extremely limited testing in real-life situations.

A user may choose to utilize these other models but must recognize their limitations after some thoughtful testing to confirm that reasonable answers are obtained in the applications of interest to the user.

PROBABILITY OF EXPLOSION

For the purposes of this tool it is assumed that the analyst either

- has access to a consequence modeling tool that will define the *potential* for an explosion to take place based on the congestion/confinement/layout present or
- assumes that an explosion is *always* possible (if there is any degree of congestion/confinement present) or is *never* possible (if the release is into an open field).

This tool therefore addresses the *probability* that an explosion takes place *given that there is explosion potential as defined by a consequence model*. Note that there are any number of reasons why an explosion would not occur, even if a cloud and the necessary congestion/confinement are present (e.g., location of the ignition source within the cloud, fuel concentration too rich to burn completely, nonidealities in the cloud dimension modeling, etc.).

The third factor that can be considered in explosion probability predictions is related to the effectiveness and availability of explosion mitigation equipment such as vent panels, deluges, etc. The design of such systems is highly situation dependent and is beyond the scope of this book. So while an option is provided in the algorithms in this book for the user to supply a probability of such systems preventing an explosion, this probability must be calculated externally based on the specific circumstances.

It is noted that Cox et al. (1990) developed a prediction for explosion probability that is not directly related to any of the factors listed above. This correlation was based on the release rate and suggests a probability of explosion given delayed ignition (POEGDI) of 0.025 for releases smaller than 1 kg/s and 0.25 for releases greater than 50 kg/s. This approach has been adopted by others and is similar to relationships developed by Ronza et al. (2007) that are related to the total amount spilled rather than the release rate. This relationship between release amount/rate and POEGDI may be reflective of two factors:

- Previous conjectures that, in the open, there is a minimum release mass required to develop an explosion.

- Large releases have a greater chance of reaching an area with the level of congestion and confinement necessary to generate explosive flame front velocities.

A few other people have characterized POEGDI proposed numbers ranging from 0.1 to 0.6, with the variations based on release orientation and weather (Crossthwaite et al., 1998) or chemical being released (API, 2000).

Level 1 Algorithm for Probability of Explosion, Given Delayed Ignition

The inputs that are important to POEGDI are not readily available to Level 1 users (such as PHA teams). Therefore, a fixed value of 0.3 is recommended. The basis for this value is somewhat vague since the physical surroundings of the release events contributing to the data are not known. However, Cox et al. (1990) developed a value of 0.25 for very large releases, and API RBI (2000) used values of 0.2–0.3 for hydrocarbons in the C3–C8 range.

Level 2 Algorithm for Probability of Explosion, Given Delayed Ignition

The factors important to POEGDI (other than congestion/confinement, which is beyond the scope of this book) that have already been obtained as part of the remaining Level 2 analysis include the following:

- Chemical involved (from a pick list; otherwise an "average" chemical is assumed)
- Magnitude of the release
- Location of release (indoors or outside).

These factors will be incorporated as described in the following sections.

Chemical Modification

The Level 1 estimate of 0.3 is the baseline for the Level 2 POEGDI modifiers. A chemical modifier M_{CHEM} will be applied as follows, based on the fundamental burning velocity (FBV):

M_{CHEM} = 0.5 if FBV is less than 45 cm/s ("low reactivity")

M_{CHEM} = 1.0 if FBV is between 45 and 75 cm/s ("medium reactivity")

M_{CHEM} = 2.0 if FBV is greater than 75 cm/s ("high reactivity")

The default value in the software is set to 1, since most materials fall in the medium-reactivity range.

Magnitude Modification

This modifier is similar in nature to the modifier described for Level 2 PODI in Section 2.8.2.3; however, some literature resources suggest that the effect of release magnitude on the POEGDI factor is not as strong as it is on the PODI factor. Therefore, the modifier is related to the PODI event magnitude modifier as follows:

$$POEGDI\ M_{MAGE} = (PODI\ M_{MAG})^{0.5} \tag{B-1}$$

For the purposes of this calculation, limits on M_{MAG} that are described in Section 2.8.2.3 of this book do not apply.

Release Location

An indoor release location will not only inhibit the dispersion of a release, but if ignited, there will be a greater propensity for explosion because of the confinement provided by the room. For this reason, a multiplier of 1.5 is applied for indoor releases ($M_{IN/OUT}$). If the release is outdoors in a process area, $M_{IN/OUT}$ = 1. If the release is outdoors in a tank farm or other remote area, set $M_{IN/OUT}$ = 0.5.

Combined Level 2 Algorithm for POEGDI

Incorporating each of the above contributors to POEGDI results in the following relationship:

$$POEGDI_{Level\ 2} = 0.3 \times M_{CHEM} \times M_{MAGE} \times M_{IN/OUT} \tag{B-2}$$

This equation must, of course, have a maximum value of 1.

Level 3 Algorithm for Probability of Explosion, Given Delayed Ignition

In most cases, the Level 2 approach to POEGDI will also be applied to Level 3. However, Level 3 will also allow the benefit of mitigative measures such as vent panels, explosion suppression systems, and the like, to be considered based on external calculations of the effectiveness and reliability of such systems. It is beyond the scope of this book to quantify the effectiveness and availability of such specialized systems for the infinite combinations of applications to which they might be applied. Therefore, the user must define the probability of failure and specify this value (FEP).

The combined Level 3 algorithm for POEGDI is then

$$POEGDI_{Level\ 3} = 0.3 \times M_{CHEM} \times M_{MAGE} \times M_{IN/OUT} \times FEP \qquad \text{(B-3)}$$

EFFECT OF VENTILATION ON THE IGNITION PROBABILITY OF INDOOR RELEASES

Introduction

Subject matter experts agree that ventilation of indoor releases can have an effect on the probability of ignition. Indeed, many buildings in chemical plants are designed to provide increased ventilation if flammable vapors are detected in the building. Therefore, there is merit in quantifying the benefit of such measures, so that they can be better defended in the safety budgeting process.

Existing Model

The only apparent published effort to quantify the effect of ventilation on ignition probability is that by Moosemiller (2010). The reasoning is somewhat lengthy but can be summarized using an "Inside Probability Multiplier" M_V (relative to outdoor releases) as follows:

$$M_V = 1.5 \times B_{es} \times B_{vr} \times B_{vdd} \qquad \text{(B-4)}$$

The individual terms are defined as follows:

1.5 = the general indoor multiplier

Table B.1. Estimating effective ventilation rate (EVR)

		Average Horizontal Distance Between Adjacent Detectors*		
		0 – 25 feet	25 – 75 feet	> 75 feet, or < 2 detectors in room
Increased Ventilation Activates Upon Detection of:	1% LFL	5 x IVR	3 x IVR	IVR
	10% LFL	3 x IVR	IVR	(IVR+NVR) / 2
	25% LFL	IVR	(IVR+NVR) / 2	(IVR x NVR)$^{1/2}$
	100% LFL	(IVR+NVR) / 2	(IVR x NVR)$^{1/2}$	NVR

IVR = Increased Ventilation Rate
NVR = Normal Ventilation Rate
LFL = Lower Flammability Limit

For flammable materials having a molecular weight of 20 or greater, include only those detectors located in the bottom 15 feet of the room/building in this calculation. If the molecular weight is less than 20, include all detectors in the building. If there are no detectors above 15 feet height for a low molecular-weight material, take the effective ventilation rate result from the table and divide by two.

Volume of Enclosed Space Factor $B_{es} = (V/150,000)^{-1/3}$, where V is the volume of the room in cubic feet. Maximum and minimum values of B_{es} are set at 3 and 0.5.

Ventilation Rate Factor $B_{vr} = (EVR/2)^{-1/2}$, where the EVR (effective ventilation rate) is expressed in air changes per hour and is related to the initial ventilation rate, any accelerated rate that may be activated upon flammable gas detection, and the location of gas detectors.

IVR and NVR in Table B.1 are also expressed in units of air changes per hour. Maximum and minimum values for B_{vr} are set at 3 and 0.3.

Ventilation Draft Direction Factor $B_{vdd} = 0.5$ if draft is designed to draw flammable gases away from likely ignition sources

$B_{vdd} = 1$ if draft is designed with no particular direction strategy in mind

$B_{vdd} = 2$ if draft is designed such that flammable gases are drawn through likely ignition sources

That is the form of the model as published. An enhancement of the above model is to ensure that the result does not give a value that is better than the outdoors would (1), at least with respect to the factors other than B_{vdd}. Therefore, it is suggested that the product $1.5 \times B_{es} \times B_{vr}$ has a minimum value of 1.

B_{vdd} may still be used as a factor in this case, on the premise that air flow that draws the flammables away from ignition sources is still beneficial and could in principle provide performance that is better than the outdoors.

Interpretation of Model for Partially Enclosed Buildings

Many process facilities have structures that house process equipment that may not be completely enclosed. Typically these will have a roof, but anywhere from zero to three walls. In principle, these areas will behave somewhere between an indoor space and an outdoor space, but possibly in ways that are not readily predictable because of the unusual air flow patterns that might result. Rather than extend the model above to account for these additional complexities, the following is suggested:

- Building with roof but no walls—use inside/outside multiplier = 1.1
- Building with roof and one wall—use inside/outside multiplier = 1.2
- Building with roof and two walls—use inside/outside multiplier = 1.3
- Building with roof and three walls—use inside/outside multiplier = 1.4

REFERENCES

API (American Petroleum Institute), "Risk-Based Inspection Base Resource Document," API Publication 581, 1st Ed., May 2000.

API (American Petroleum Institute), "Recommended Practice for Classification of Locations for Electrical Installation at Petroleum Facilities Classified as Class I, Division 1 and Division 2," API Recommended Practice 500, 2002.

API (American Petroleum Institute), "Ignition Risk of Hydrocarbon Liquids and Vapors by Hot Surfaces in the Open Air," API Recommended Practice 2216, 3rd Ed.," December 2003.

API (American Petroleum Institute), "Spark Ignition Properties of Hand Tools," API Recommended Practice 2214, 4th Ed., July 2004.

API (American Petroleum Institute), "Risk Based Inspection Technology," API Recommended Practice 581, 2008.

API (American Petroleum Institute), "Management of Hazards Associated with Location of Process Plant Permanent Buildings," API Recommended Practice 752, 3rd Ed., December 2009.

ASTM International, ASTM Computer Program for Chemical Thermodynamic and Energy Release Evaluation CHETAH Version 9.0 - DS51F, 2011.

Babrauskas, V., "Ignition Handbook," Fire Science Publishers/SFPE, 2003.

Bragin, M.V. and V.V. Molkov, International Journal of Hydrogen Energy, Volume 36, Issue 3, pages 2589–2596, February 2011.

British Standards (BS), "Explosive Atmospheres—Classification of Areas," BS EN 60079, Part 10-1 (2009).

Britton, L.G., D.A. Taylor, and D.C. Wobser, "Thermal Stability of Ethylene at Elevated Pressures," Plant/Operations Progress, 5: 238–251, 1986.

Britton, L.G., "Combustion Hazards of Silane and Its Chlorides," Plant/Operations Progress, Vol. 9, No. 1, January 1990(a).

Britton, L.G., "Thermal Stability and Deflagration of Ethylene Oxide," Plant/Operations Progress, Vol. 9, No. 2, April 1990(b).

Britton, L.G., "Avoiding Static Ignition Hazards in Chemical Operations," Center for Chemical Process Safety, American Institute of Chemical Engineers, 1999.

Britton, L.G., "Using Heats of Oxidation to Evaluate Flammability Hazards," Process Safety Progress, Vol. 21, No. 1, March 2002.

Britton, L.G. and J.A. Smith, "Static Hazards of the VAST," Journal of Loss Prevention in the Process Industries. 25 (2012), 309-328.

Catoire, L. and V. Naudet, "A Unique Equation to Estimate Flash Points of Selected Pure Liquids—Application to the Correction of Probably Erroneous

Flash Point Values," Journal of Physical and Chemical Reference Data, Vol. 33, No. 4, 2004.

Cawley, J., "Probability of Spark Ignition in Intrinsically Safe Circuits," Bureau of Mines Report of Investigations RI 9183, 1988.

CCPS, "Guidelines for Engineering Design for Process Safety," Center for Chemical Process Safety/American Institute of Chemical Engineers, New York, 1993.

CCPS, "Guidelines for Post-release Mitigation Technology in the Chemical Process Industry," Center for Chemical Process Safety/American Institute of Chemical Engineers, New York, 1997.

CCPS, "Guidelines for Chemical Process Quantitative Risk Analysis," 2nd Ed., Center for Chemical Process Safety/American Institute of Chemical Engineers, New York, 1999.

CCPS, "Guidelines for Vapor Cloud Explosion, Pressure Vessel Burst, BLEVE and Flash Fire Hazards," 2nd Ed., Center for Chemical Process Safety/American Institute of Chemical Engineers, New York, 2010.

CCPS, "Guidelines for Engineering Design for Process Safety," 2nd Ed., Center for Chemical Process Safety/American Institute of Chemical Engineers, New York, 2012(a).

CCPS, "Guidelines for Evaluating Process Plant Buildings for External Explosions, Fires, and Toxic Releases," 2nd Ed., Center for Chemical Process Safety/American Institute of Chemical Engineers, New York, 2012(b).

CCPS, "Guidelines for Enabling Conditions and Conditional Modifiers for Layer of Protection Analysis," Center for Chemical Process Safety/American Institute of Chemical Engineers, New York, 2013.

Cox, A.W., Lees, F.P. and M.L. Ang, "Classification of Hazardous Locations," IChemE, 1990.

Crossthwaite, P.J., Fitzpatrick, R.D. and N.W. Hurst, "Risk Assessment for the Siting of Developments Near Liquefied Petroleum Gas Installations," IChemE Symposium Series No. 110, 1988.

Crowl, D.A., "Understanding Explosions," Center for Chemical Process Safety/American Institute of Chemical Engineers, New York, 2003.

Daycock, J.H., and P.J. Rew, "Development of a Method for the Determination Of On-Site Ignition Probabilities," Health & Safety Executive Research Report 226, 2004.

Dryer, F., Chaos, M., Zhao, Z., Stein, J., Alpert, J. and C. Homer, "Spontaneous Ignition of Pressurized Releases of Hydrogen and Natural Gas into Air," Combustion Science and Technology, Vol. 179, pages 663–694, 2007.

Duarte, D., Rohalgi, J. and R. Judice, "The Influence of the Geometry of the Hot Surfaces on the Autoignition of Vapor/Air Mixtures: Some Experimental and Theoretical Results," Process Safety Progress, Vol. 17, Spring 1998.

E&P Forum, "Risk Assessment Data Directory," Report No. 11.8/250, International Association of Oil & Gas Producers, October, 1996.

Foster, K.J. and J.D. Andrews, "Techniques for Modeling the Frequency of Explosions on Offshore Platforms," Proceedings of the Institution of Mechanical Engineers., Vol. 213, Part E, pages 111–119, IMechE, 1999.

Fthenakis, V., Ed., "Prevention and Control of Accidental Releases of Hazardous Gases," Van Nostrand Reinhold, New York, 1993.

Garland, R.W., "Quantitative Risk Assessment Case Study for Organic Acid Processes," Process Safety Progress, Vol. 29, No. 3, September 2010.

Glor, M., "Electrostatic Ignition Hazards Associated with Flammable Substances in the Form of Gases, Vapors, Mists and Dusts," Institute of Physics Conference Series. No. 163, March 1999.

Gummer, J. and S. Hawksworth, "Spontaneous Ignition of Hydrogen," HSE Books, Research Report RR615, 2008.

Hamer, P.S., Wood, B.M., Doughty, R.L., Gravell, R.L., Hasty, R.C., Wallace, S.E. and J.P. Tsao, "Flammable Vapor Ignition Initiated by Hot Rotor Surfaces Within an Induction Motor: Reality or Not?" IEEE Transactions on Industry Applications, Vol. 35, No. 1, January/February 1999.

HMSO/UK Health & Safety Executive, "Canvey—A Second Report. A Review of Potential Hazards from Operations in the Canvey Island/Thurrock Area Three Years after Publication of the Canvey Report," 1981.

Hooker, P., Royle, M., Gummer, J., Willoughby, D. and J. Udensi, Hazards XXII, Symposium Series No. 156, pages 432–439, 2011.

HSE (Health and Safety Executive) website, http://www.buncefieldinvestigation.gov.uk/reports/index.htm, 2012.

IEC 60079-32-1/TS/Ed1, "Explosive Atmospheres—Part 32-1: Electrostatic Hazards, Guidance," http://www.tk403.ru/pdf/pdf_inter/31_1033e_DTS.pdf, accessed January 2013.

ISO, "Fire Safety—Vocabulary (ISO 13943)," International Organization for Standardization, Geneva, 2008.

Jallais, S., "Hydrogen Ignition Probabilities," internal presentation, August, 2010.

Johnson, R.W., "Ignition of Flammable Vapors by Human Electrostatic Discharges," AIChE 14th Loss Prevention Symposium, June 8–12 1980, Philadelphia, PA.

Klinkenberg, A. and van der Minne, J., Eds., "Electrostatics in the Petroleum Industry—The Prevention of Explosion Hazards," Elsevier Publishing Co., 1958.

Kuchta, J.M., "Investigation of Fire and Explosion Accidents in the Chemical, Mining, and Fuel-Related Industries," Bulletin No. 680, U.S. Bureau of Mines, 1985.

Lee, K-P., Wang, S-H. and S-C. Wong, "Spark Ignition Characteristics of Monodisperse Multicomponent Fuel Sprays," Combustion Science and Technology, Vol. 113-4, 1996.

Liao, C., Terao, K. and Y. Utaka, "Ignition Probability in a Fuel Spray," Japanese Journal of Applied Physics, Vol. 31, No. 7, July 1992.

Mannan, S, Ed., "Lees' Loss Prevention in the Process Industries," 3rd Ed., Butterworth-Heinemann, 2005.

Molkov, V., "Hydrogen Safety Research: State-of-the-Art," Proceedings of the 5th International Seminar on Fire and Explosion Hazards, Edinburgh, UK, April 23–27, 2007.

Moosemiller, M.D., "Development of Algorithms for Predicting Ignition Probabilities and Explosion Frequencies," Process Safety Progress, Vol. 29, No. 2, June 2010.

Murphy, J., "Remote Isolation of Process Equipment," http://www.aiche.org/uploadedFiles/CCPS/Resources/KnowledgeBase/Final%20Remote%20Isolation%20Aug09.pdf.

NFPA, "NFPA 1—Fire Code," National Fire Protection Association, Quincy, MA, 2012.

NFPA, "NFPA 68: Standard on Explosion Protection by Deflagration Venting," Quincy, MA, 2013.

NFPA, "NFPA 77: Recommended Practice on Static Electricity, 2014 Edition," Quincy, MA, 2013.

OSHS, "Guidelines for the Control of Static Electricity in Industry," Occupational Safety and Health Service, Department of Labour, Wellington, New Zealand, 1982 (rev. 1990, internet version 1999).

Pesce, M., Paci, P., Garrone, S., Pastorino, R. and B. Fabiano, "Modeling Ignition Probabilities within the Framework of Quantitative Risk Assessments," Chemical Engineering Transactions, Vol. 26, 2012.

Pratt, T.H., "Electrostatic Ignitions of Fires and Explosions," Center for Chemical Process Safety, American Institute of Chemical Engineers, 2000.

RIVM (National Institute of Public Health and the Environment), "Reference Manual Bevi Risk Assessments," Version 3.2, Bilthoven, Netherlands, 2009.

Ronza, A., Vílchez, J. and J. Casal, "Using Transportation Accident Databases to Investigate Ignition and Explosion Probabilities of Flammable Spills," Journal of Hazardous Materials, Vol. 146, pages 106–123, 2007.

Rowley, J.R., Rowley, R.L. and W.V. Wilding, "Estimation of the Flash Point of Pure Organic Chemicals from Structural Contributions," Process Safety Progress, Vol. 29, No. 4, December 2010.

SFPE (Society of Fire Protection Engineers), "Handbook of Fire Protection Engineering," 4th Ed., Society of Fire Protection Engineers/National Fire Protection Association, 1998.

SFPE (Society of Fire Protection Engineers), "SFPE Handbook of Fire Protection Engineering," 4th Ed., National Fire Protection Association, 2008.

Smith, D., "Study Indicates Risk to LDC Assets Posed by Static Electricity," Pipeline Gas Journal, Vol. 238, No. 4, April 2011.

Spencer, H. and P.J. Rew, "Ignition Probability of Flammable Gases," Health & Safety Executive Contract Research Report 146, 1997.

Spencer, H., Daycock, J. and P.J. Rew, "A Model for the Ignition Probability of Flammable Gases, Phase 2," Health & Safety Executive Contract Research Report 203, 1998.

Spouge, J., "A Guide to Quantitative Risk Assessment for Offshore Operations," CMPT Publication 99/100, 1999.

Srekl, J. and J. Golob, "New Approach to Calculate the Probability of Ignition," presented at 8thWorld Congress of Chemical Engineering, Montreal, Quebec, Canada, August 23–27, 2009,

Swain, M.R., Filoso, P.A. and M.N. Swain, "An Experimental Investigation into the Ignition of Leaking Hydrogen," International Journal of Hydrogen Energy, Vol. 32, No. 2, pages 287–295, 2007.

Thomas, J.K., Kolbe, M., Goodrich, M.L. and E. Salzano, "Elevated Internal Pressures in Vented Deflagration Tests," 40th Annual Loss Prevention Symposium, Orlando, FL, April 24–27, 2006.

Thyer, A.M., "Offshore Ignition Probability Arguments," Report Number HSL/2005/50, Health and Safety Laboratory, U.K., 2005.

TNO Purple Book, prepared for the Committee for the Prevention of Disasters, "Guidelines for Quantitative Risk Assessment, CPR18E. SDU," The Hague, 2005.

Tromans, P.S. and R.M. Furzeland, "An Analysis of Lewis Number and Flow Effects on the Ignition of Premixed Gases," 21st Symposium (International) on Combustion/The Combustion Institute, pages 1891–1897, 1986.

UKEI, "Ignition Probability Review, Model Development and Look-Up Correlations," Energy Institute, London, 2006.

Wehe, S.D. and N. Ashgriz, "Ignition Probability and Absolute Minimum Ignition Energy in Fuel Sprays," Combustion Science and Technology, Vol. 86, No. 1, 1992.

Witcofski, R.D., "Dispersion of Flammable Clouds Resulting from Large Spills of Liquid Hydrogen," National Aeronautics and Space Administration Report No. NASA TM-83131, May 1981.

WOAD (World Offshore Accident Data), Det Norske Veritas, 1994.

Wolanski, P. and S. Wojcicki, "Investigation into the Mechanism of the Diffusion Ignition of a Combustible Gas Flowing into an Oxidizing Atmosphere," Proceedings of the Combustion Institute, Vol.14, pp. 1217-1223, 1972.

Zabetakis, M.G., "Flammability Characteristics of Combustible Gases and Vapors," U.S. Bureau of Mines Bulletin 627, 1965.

Zalosh, R., Short, T., Marlin, P. and D. Coughlin, "Comparative Analysis of Hydrogen Fire and Explosion Incidents," Progress Report No. 3, for Division of Operational and Environmental Safety, U.S. Department of Energy, Contract No. EE-77-C-02-4442, July 1978.

INDEX

Printed and bound by CPI Group (UK) Ltd, Croydon, CR0 4YY

23/04/2025

14660907-0002